NATO ASI Series

Advanced Science Institutes Series

A series presenting the results of activities sponsored by the NATO Science Committee, which aims at the dissemination of advanced scientific and technological knowledge, with a view to strengthening links between scientific communities.

The Series is published by an international board of publishers in conjunction with the NATO Scientific Affairs Division

A Life Sciences **B Physics**	Plenum Publishing Corporation London and New York
C Mathematical and Physical Sciences **D Behavioural and Social Sciences** **E Applied Sciences**	Kluwer Academic Publishers Dordrecht, Boston and London
F Computer and Systems Sciences **G Ecological Sciences** **H Cell Biology** **I Global Environmental Change**	Springer-Verlag Berlin Heidelberg New York London Paris Tokyo Hong Kong Barcelona Budapest

NATO-PCO DATABASE

The electronic index to the NATO ASI Series provides full bibliographical references (with keywords and/or abstracts) to more than 30000 contributions from international scientists published in all sections of the NATO ASI Series. Access to the NATO-PCO DATABASE compiled by the NATO Publication Coordination Office is possible in two ways:

- via online FILE 128 (NATO-PCO DATABASE) hosted by ESRIN, Via Galileo Galilei, I-00044 Frascati, Italy.
- via CD-ROM „NATO Science & Technology Disk“ with user-friendly retrieval software in English, French and German (© WTV GmbH and DATAWARE Technologies Inc. 1989).

The CD-ROM can be ordered through any member of the Board of Publishers or through NATO-PCO, Overijse, Belgium.

Series I: Global Environmental Change, Vol. 6

The ASI Series Books Published as a Result of Activities of the Special Programme on Global Environmental Change

This book contains the proceedings of a NATO Advanced Research Workshop held within the activities of the NATO Special Programme on Global Environmental Change, which started in 1991 under the auspices of the NATO Science Committee.

The volumes published as a result of the activities of the Special Programme are:

Vol. 1: **Global Environmental Change.** Edited by R. W. Corell and P. A. Anderson. 1991.
Vol. 2: **The Last Deglaciation: Absolute and Radiocarbon Chronologies.** Edited by E. Bard and W. S. Broecker. 1992.
Vol. 3: **Start of a Glacial.** Edited by G. J. Kukla and E. Went. 1992.
Vol. 4: **Interactions of C, N, P and S Biogeochemical Cycles and Global Change.** Edited by R. Wollast, F. T. Mackenzie and L. Chou. 1993.
Vol. 5: **Energy and Water Cycles in the Climate System.** Edited by E. Raschke and D. Jacob. 1993.
Vol. 6: **Prediction of Interannual Climate Variations.** Edited by J. Shukla. 1993.
Vol. 7: **The Tropospheric Chemistry of Ozone in the Polar Regions.** Edited by H. Niki and K. H. Becker. 1993.

Prediction of Interannual Climate Variations

Edited by

J. Shukla

Center for Ocean-Land-Atmosphere Interactions
Department of Meteorology
University of Maryland at College Park
College Park, MD 20742-2425, USA

Springer-Verlag
Berlin Heidelberg New York London Paris Tokyo
Hong Kong Barcelona Budapest
Published in cooperation with NATO Scientific Affairs Division

Proceedings of the NATO Advanced Research Workshop on Prediction of Interannual Climate Variations held at Trieste, Italy, July 22–26, 1991

ISBN-13: 978-3-642-76962-7 e-ISBN-13: 978-3-642-76960-3
DOI: 10.1007/978-3-642-76960-3

Softcover reprint of the hardcover 1st edition 1993

Typesetting: Camera ready by authors
31/3145 - 5 4 3 2 1 0 - Printed on acid-free paper

TABLE OF CONTENTS

PART 3: Low-Frequency Variability

PART 4: Present Status of Seasonal Forecasting

PART 5: Predictability

ACKNOWLEDGEMENT

This workshop witnessed a gathering of the most distinguished key world leaders in the prediction and predictability of weather and climate. Directors and senior researchers of the world's leading institutions in climate modeling and prediction research were on hand and evoked stimulating discussions on the current state of knowledge and directions for future research. We are grateful to all the scientists for their participation and their excellent presentations.

Special thanks are extended to the members of the organizing committee for their planning of the workshop: Dr. L. Bengtsson, Dr. D. Carson, Professor G. Furlan and Professor P. Morel.

The excellent local arrangements for the workshop and the social programs in Trieste were made by Ms. Lisa Iannitti. We are grateful to Lisa for her dedication and efficiency, and above all, for her ability to keep a cheerful atmosphere in spite of above normal temperature and humidity.

Finally, we are grateful to NATO for their encouragement and support of the emerging area of global environmental change.

LIST OF FIRST NAMED AUTHORS

Dr. Phil Arkin
Office of Global Programs
NOAA R/CAR, Room 4200
1355 East–West Highway
Silver Spring, MD 20910 USA

Dr. Michael Davey
Robert Hooke Institute
Clarendon Laboratory
Parks Road
Oxford OX1 3PU
UNITED KINGDOM

Mr. Paul A. Dirmeyer
Center for Ocean–Land–Atmosphere Interactions
Department of Meteorology
University of Maryland
College Park, MD 20742 USA

Ms. Lydia Dumenil
Max–Planck–Institut fur Meteorologie
Bundesstrafe 55
D–2000 Hamburg 13
GERMANY

Dr. Kikuro Miyakoda
Geophysical Fluid Dynamics Laboratory
P.O. Box 308
Princeton University
Princeton, NJ 08542 USA

Dr. J. F. Royer
Centre National de Recherches Meteorologiques
42 Av. Coriolis
F–31057 Toulouse Cedex
FRANCE

Dr. J. Servain
Centre OSTOM/IFREMER
B. P. 337
29273 Brest Cedex
FRANCE

Dr. J. Shukla
Center for Ocean–Land–Atmosphere Interactions
University of Maryland
College Park, MD 20742 USA

Dr. Antonio Speranza
Associazione per lo Studio e le sul Futuro del Sistema interattive Uomo–Terra
Via M. Dionigi, 29
00193 Rome ITALY

Dr. S. Tibaldi
Dept. of Physica, Atmospheric Dynamics Group
University of Bologna
Via Irnerio, 46
40126 Bologna, ITALY

Dr. J. Tribbia
National Center for Atmospheric Research
P.O. Box 3000
1850 Table Mesa Drive
Boulder, CO 80307 USA

Dr. John E. Walsh
Dept. of Atmospheric Sciences
University of Illinois
Urbana, IL 61801 USA

Dr. M. N. Ward
Hadley Climate Center
Meteorological Office
London Road
Bracknell, RG122SZ
UNITED KINGDOM

LIST OF OTHER PARTICIPANTS

Dr. Lennart Bengtsson
Max–Planck–Institut fur Meteorologie
Bundesstrafe 55
D–2000 Hamburg 13
GERMANY

Dr. David Burridge
European Centre for Medium Range
Weather Forecasts
Shinfield Park
Reading, Berkshire RG2 9AX
UNITED KINGDOM

Dr. Mark Cane
Lamont–Doherty Geological Observatory
Columbia University
Palisades, NY 10964 USA

Dr. D. Carson
Hadley Climate Center
Meteorological Office
London Road
Bracknell, RG122SZ
UNITED KINGDOM

Prof. M. Graziani
International Center for Science & High Technology
P.O. Box 586
Adriatico Palace, Via Grignano, 9
34100 Trieste, ITALY

Prof. G. Furlan
International Centre for Theoretical Physics
P.O. Box 586, Miramare
Strada Costieva 11
34100 Trieste
ITALY

Dr. Michael J. Hall
Director, Office of Global Programs, NOAA
Room 4200
1355 East–West Highway
Silver Spring, MD 20910 USA

Dr. M. Latif
Max Planck Institute fur Meteorologie
Bundestrasse 55
D–2000 Hamburg, GERMANY 13

Dr. Roberto Legnani
IMGA–CNR, via Emilia Est. 770
41100 Modena, ITALY

Dr. R. McPherson
National Meteorological Center
W32 World Weather Building, Room 204
5200 Auth Road
Washington, DC 20233 USA

Dr. Kenneth Mooney
U.S. TOGA Project Office
NOAA R/CAR, Room 4200
1355 East–West Highway
Silver Spring, MD 20910 USA

Pierre Morel
World Climate Research Programme
World Meteorological Organization
41 avenue Giuseppe–Motta
C.P. No. 2300
1211 Geneva 2
SWITZERLAND

Dr. Antonio D. Moura
Director of Meteorology
INPE – C. Postal 515
12201 – Sao Jose dos Campos – SP
BRAZIL

Dr. N. Nigro
International Centre for Science and High Technology
P.O. Box 586
via Beirut 7
Adriatico Palace, Via Grignano, 9
34100 Trieste, ITALY

Dr. Eric Pitcher
Cray Research Incorporated
133 Northland Drive
Mendota Heights, MN 55120 USA

Mr. Michael Probonas
Taxilou 29
157 71w Zografou
Athens, GREECE

Dr. S. Radicella
International Centre for Theoretical Physics
P.O. Box 586
Via Beirut 7
34100 Trieste, ITALY

Dr. Anna Rampa
via Rubens 8
20148 Milano ITALY

Dr. Robert Sadourny
Laboratoire de Meteorologie Dynamique
Ecole Normale Superieure
24 rue Lhomond
75231 Paris Decex 05
FRANCE

Dr. Abdus Salam
Director, International Centre for Theoretical Physics
P.O. Box 586, Miramare
Strada Costieva 11
34100 Trieste, ITALY

Dr. Ed Sarachik
Department of Atmospheric Sciences
(AK–40)
University of Washington
Seattle, WA 98195 USA

Dr. Edwin K. Schneider
Center for Ocean–Land–Atmosphere Interactions
Department of Meteorology
University of Maryland
College Park, MD 20742 USA

Dr. A. Stravisi
International Centre for Theoretical Physics
P.O. Box 586
Via Beirut 7
34100 Trieste, ITALY

Dr. A. Sun
International Centre for Science & High Technology
P.O. Box 586
via Beirut 7
34100 Trieste, ITALY

Dr. Alfonso Sutera
Department of Math & Physics
University di Camerino
Via Madonna delle Carceri
62032 Camerino (MC)
ITALY

Prof. E. Tosatti
International School for Advanced Studies (ISAS)
via Beirut 2
34100 Trieste, ITALY

PREFACE

It has been known for some time that the behavior of the short-term fluctuations of the earth's atmosphere resembles that of a chaotic non-linear dynamical system, and that the day-to-day weather cannot be predicted beyond a few weeks. However, it has also been found that the interactions of the atmosphere with the underlying oceans and the land surfaces can produce fluctuations whose time scales are much longer than the limits of deterministic prediction of weather. It is, therefore, natural to ask whether it is possible that the seasonal and longer time averages of climate fluctuations can be predicted with sufficient skill to be beneficial for social and economic applications, even though the details of the day-to-day weather cannot be predicted beyond a few weeks.

The main objective of the workshop was to address this question by assessing the current state of knowledge on predictability of seasonal and interannual climate variability and to investigate various possibilities for its prediction.

The workshop participants generally agreed that it should be possible to predict the seasonal and longer time averaged atmospheric circulation. This expectation is based on a large number of observational and modeling research projects during the last 15-20 years, which shows that almost all the major interannual climate fluctuations observed during the past 100 years of observational records were caused either by atmosphere-ocean, or atmosphere-ocean-biosphere interactions. Recently, our ability to simulate the earth's climate using high speed computers has also improved significantly. It is, therefore, quite timely to ask: Have we reached a stage where an organized and coordinated effort can be launched to begin to make routine dynamical predictions of short-term climate variability? The workshop explored this question by reviewing the scientific progress and exploring various possibilities to make climate prediction.

In a series of comprehensive review papers, the relationships between changes in atmospheric circulations and changes in sea surface temperature, soil moisture, sea ice and snow were firmly established. The influence of each of these boundary conditions was reviewed separately based on analysis of past observations and sensitivity experiments with global climate models. Tropical sea surface temperature anomalies have a profound effect on tropical circulation and rainfall, and in some cases, also influence the extra-tropical atmospheric circulation. Several recent studies have also shown a large and significant correlation between atmospheric circulation and extra-tropical sea surface temperature anomalies. There are also several examples of significant correlations between winter snow cover and atmospheric circulation anomalies in the subsequent season, the influence of initial soil moisture on subsequent atmosphere circulation up to one season, and the influence of sea ice anomalies on atmospheric circulation. It was, therefore, concluded that for monthly and seasonal predictions, it is extremely important that the observed initial state of the atmosphere and the boundary conditions of sea surface temperature, soil wetness, sea ice, snow depth and snow cover be mutually consistent with each other.

It was also recognized that for predictions beyond a season, and especially for predictions of interannual variability, it is necessary to predict the boundary conditions at the earth's surface. Significant progress has already been made by demonstrating the feasibility of predicting tropical sea surface temperature fluctuations using coupled models of the tropical oceans and atmosphere.

Although there are considerable deficiencies in the currently available coupled ocean-atmosphere models, it was generally agreed that the best strategy to make advances in prediction of interannual climate variations is to employ state of the art general circulation models of the oceans, the atmosphere and the land surface processes. This will require development of coupled ocean-atmosphere models which can realistically simulate the observed annual cycle and interannual variations. This will also require the development of data assimilation and initialization techniques for coupled models.

It was also recognized that modelling and prediction of regional and interannual climate variations is important for quantifying the degree of uncertainty in predicting global climate changes due to human activities.

J. SHUKLA

OBSERVATIONAL AND MODELING STUDIES OF THE INFLUENCE OF SOIL MOISTURE ANOMALIES ON ATMOSPHERIC CIRCULATION (REVIEW)

P. A. Dirmeyer and J. Shukla
Center for Ocean-Land-Atmosphere Interactions
Department of Meteorology
University of Maryland
College Park, Maryland 20742-2425
USA

1. Introduction

The annual mean rainfall for the global continents is estimated to be about 764 mm of which approximately 40% runs off into the oceans. Assuming no secular trends in the annual mean global soil moisture, this suggests that the annual and global mean evaporation from the land surfaces alone is about 60% of the annual and global mean precipitation over the land (Eagleson, 1991). The percentage is even higher during the local summer and in the tropics.

Some of the evaporated moisture may condense and reprecipitate locally. One region where this local water cycle is strong is northern South America. Several independent water balance calculations for the Amazon basin show that approximately 55% of the precipitation there is accounted for by local evapotranspiration (see Salati and Nobre, 1991 for review). On the other side hand, about 75% of the precipitation which falls on the basin is reevaporated before reaching the oceans as runoff (Dickinson, 1991). Figure 1 shows the hydrologic cycle over the Amazon basin.

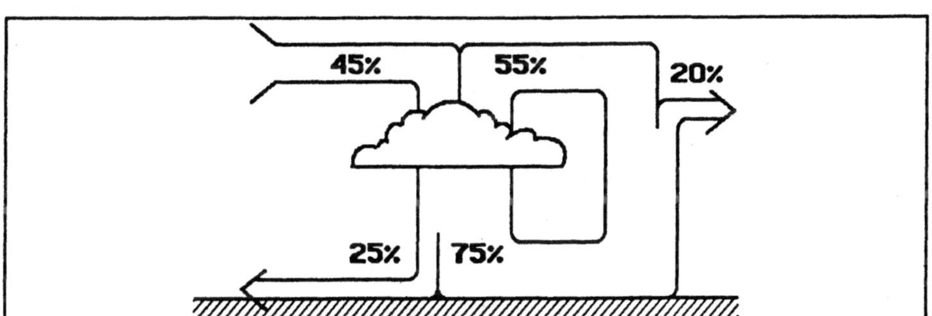

Figure 1. Hydrologic cycle over the Amazon Basin.

NATO ASI Series, Vol. I 6
Prediction of Interannual Climate Variations
Edited by J. Shukla

Evaporation from the land surface is a very important component of the global water budget and hydrological cycle. However, it does not necessarily follow that the water evaporated from the land is important in determining the rainfall over land. For example, all the water evaporated from the land could be advected away to the oceans before it recondenses and rains. In that case it will affect the moisture budget and evaporation only over the oceans, which in turn will, of course, affect the moisture supply for rainfall over the land. In order that evaporation from the land affects the rainfall over the land, it is necessary that the prevailing dynamical circulation be such that the land-evaporated moisture recondenses as rain before being advected away. This will depend upon the geographical location of the region under consideration, the prevailing advective velocity, the structure and intensity of the convergence field, and the vertical distribution of moist static energy which determines the nature of the moist convection.

In nature, the total rainfall averaged over the appropriate space and time scales is determined by the combined effects of available and precipitable moisture, and the character and intensity of the dynamical circulation necessary to lift the moisture for condensation and precipitation.

The role of soil moisture is twofold. First, it determines the rate of evaporation, and therefore the moisture supply. Second, it influences the partitioning of incoming radiative energy into sensible and latent heating. Soil wetness influences the heating of the ground which determines the sensible heat flux and affects the dynamical circulation by generation or dissipation of "heat lows". The interaction between the heat lows generated by solar heating of the ground and the associated circulation and rainfall is further complicated by the fact that the maintenance and the intensification of the dynamical low is largely influenced by the latent heat of condensation. For example, if the soil is saturated with water, and the evaporation is equal to the potential evaporation, there will be maximum possible supply of moisture to the atmosphere. Whether this will increase the rainfall or not will depend upon the nature of dynamical circulation and its associated flow patterns. If the rate at which moist static energy is advected away from the region is larger than its accumulation rate, it will not lead to any increase in rainfall. In the reverse case, it will. For the other extreme situation, when the soil is completely dry and there is no evaporation from the land, there may be a reduction in the rainfall. However, if the heating of the land produced intense low pressure areas which can converge moisture from the surrounding oceans, the rainfall may not necessarily decrease, and if the convergence of moisture is large enough it may even increase the rainfall. This mechanism will cease to operate once the rain starts falling because the soil will no longer be dry.

When the soil does contain water, it acts as a shallow, but widespread, reservoir. Compared with the oceans, it is a highly variable and inconsistent source of moisture for the atmosphere. The rate of evaporation depends not only on the availability of moisture in the uppermost layers of the soil, but on characteristics of the soil itself, the type and distribution of vegetation rooted in the soil, and ambient conditions in the atmosphere near the surface of the earth.

Soil moisture is an important link in two of the principal cycles or feedback loops of the earth climate system — the water cycle and the energy cycle. Perhaps most apparent is the role the land surface holds in the hydrologic cycle. Moisture evaporates from the soil, increasing atmospheric humidity and eventually condensing into clouds where it may precipitate back onto the earth's surface. Less obvious is the role soil moisture plays in the energy cycle. The evaporation of soil moisture constitutes a flux of latent heat into the atmosphere. Thus, the availability of soil moisture is a strong control on temperature and the partitioning of energy at the surface. Changes in the energy balance affect atmospheric temperature, and thus, a host of other components of the climate including evaporation itself.

In this review, we will examine some observational evidence that soil moisture fluctuations do indeed affect climate over seasonal time scales. A few of the many computer modeling studies also will be reviewed. Special attention will be given to those studies which explore the use of soil moisture in the prediction of climate.

2. Observational Studies

In this section we review the observational evidence that soil moisture fluctuations affect atmospheric circulation and rainfall at seasonal time scales.

2.1 Sensitivity to soil moisture

Although the importance of soil moisture to the generation of precipitation was stated at least as early as 1935 by E. P. Stebbing (Anthes, 1984), Namias (1959, 1960) was perhaps the first to directly address the problem of soil wetness as a boundary forcing for the atmosphere. He examined monthly precipitation and temperature data for the Great Plains of the United States. The data were seasonally averaged over periods of 60-84 years, depending on availability of the data. Namias then constructed contingency tables with three categories each (normal, above normal and below normal). Tables relating summer temperature to both temperature and precipitation from the antecedent spring show a clear tendency for dry springs to be followed by hot summers, and wet springs to be followed by cool summers (see Table 1). Also, a tendency for persistence of anomalous temperatures from spring to summer was evident.

Table 1.

	Western Plains	Subsequent summer temperature			
Spring Temp.	Precip.	Cold	Normal	Warm	Total
Cold		*101*	*70*	*40*	
	Light	29	21	10	*60*
	Moderate	31	18	19	*67*
	Heavy	**41**	31	11	*83*
Normal		*53*	*74*	*81*	
	Light	12	18	34	*64*
	Moderate	18	33	27	*78*
	Heavy	23	23	19	*65*
Warm		*57*	*65*	*87*	
	Light	9	27	**50**	*86*
	Moderate	18	22	22	*62*
	Heavy	30	16	16	*62*

from Namias (1960)

Walsh et al. (1985) have found that errors in surface temperature forecasts for the United States during 1947-80 correlated well to soil moisture anomalies. The mean errors of temperature vary by 0.5-0.7°C over most of the Great Plains and Rocky Mountains according to the sign of the anomaly of soil moisture. About half of the 61 stations included in the study had differences in composited mean specification which were statistically significant at the 95% confidence level for the summer months when errors were categorized according to the sign of the soil moisture anomaly.

Namias (1960) found that for summer precipitation, just as wet/dry springs tend to precede cool/warm summers, cool/warm springs precede wet/dry summers (see Table 2). Especially strong was the tendency for a warm dry spring to usher in a dry summer. Barnston and Schickcdanz (1984) found statistical evidence that on a smaller scale, irrigation may increase precipitation. This seems to be especially true when there was low-level mesoscale convergence over the irrigated area.

Table 2.

	Western Plains	Subsequent summer precipitation		
Spring temp.	Precip.	Below normal	Near normal	Above normal
Cold		*53*	*73*	*85*
	Light	12	18	30
	Moderate	19	24	25
	Heavy	22	31	30
Normal		*70*	*73*	*65*
	Light	28	17	20
	Moderate	27	26	26
	Heavy	15	30	19
Warm		*87*	*63*	*58*
	Light	49	22	14
	Moderate	24	16	22
	Heavy	14	25	22

based on Namias (1960)

Namias concluded that, "...moist soil may serve as a cooling reservoir by using for vaporization some of the heat normally associated with the spring to summer building of the upper level anticyclone...," (Namias, 1959), and "...desiccating warm and dry weather over the Plains in spring provides a healthy environment for the lodgement of the upper level anticyclone in the following summer," (Namias, 1960). More recently, Namias (1989) asserted that low springtime soil moisture was a factor in the US drought of 1988, and a recent study by Fennessy and Shukla (1992) supports this possibility. Any effects that soil moisture would have on the atmosphere should occur at the decay time scale of soil moisture. This has been found by Vinnikov and Yeserkepova (1991) to be typically 2-3 months for a 1 m depth of soil. Modeling studies by Carson and Sangster (1981), Rind (1982) and Delworth and Manabe (1988), among others, have implied similar decay time scales.

2.2 Soil moisture as a predictor

Namias recognized that soil moisture anomalies could aid in the persistence of atmospheric circulation anomalies. In particular he examined two case studies (Namias,1959), one involving heavy spring rains over Texas, and one drought over the eastern United States. In the first case, it was found that from February through mid-June, statistical predictions of 700 mb height for consecutive two-week periods were consistently too high over the region. He associated the anomalous trough over the area to the moist soil acting as a cooling reservoir which impeded the normal building of the summertime upper-level anticyclone in that area. In the instance of the drought over the eastern seaboard, July surface temperatures were abnormally high, even though the region was under an area of anomalously low heights and cold advection at 700 mb.

Can soil moisture be used effectively as a prognostic tool? Karl (1983, 1986) has shown that soil moisture indices can aid in long-range forecasting, particularly in spring and early summer. Karl inspected monthly averaged temperature and precipitation for the United States from 1895-1981. From this he computed the Palmer Drought Severity Index (PDSI) (Karl, 1983), moisture anomaly index Z and water content parameter WC (Karl, 1986). All are derived from the Palmer Drought Model (Palmer, 1965). The moisture anomaly index is where K is a spatially and temporally

$$Z = (P - \hat{P})K \ ,$$

dependant weighting factor used to standardize the index. P is actual precipitation, and

$$\hat{P} = \hat{ET} + \hat{R} + \hat{RO} - \hat{L}$$

The circumflex indicates climatically appropriate quantities for the existing conditions. *ET* is evapotranspiration, *R* is soil moisture recharge, *RO* is runoff, and *L* is monthly soil moisture loss. The water content parameter is computed as:

$$WC = W_s + W_u \ .$$

where *s* denotes the surface soil layer, and *u* is the underlying layer.

Sensitivity tests and contingency tables suggested that there is more persistence in rainfall in the Rocky Mountain and Great Plains states than in areas where moist advection from oceans is present. Seasonal *Z* and *WC* correlate to subsequent seasonal temperature better than straight persistence of seasonal temperature. This was especially true in the continental interior. Karl decided that PDSI is not a good predictor of rainfall, perhaps because PDSI is not very sensitive to real changes in soil moisture. *Z* and *WC* show skill as predictors, but are still very sensitive to the method of calculating evapotranspiration.

Soil moisture may have usefulness as a predictor outside the interests of atmospheric science. Serafini and Sud (1982) developed a model for the calculation of agricultural drought inception time as a function of soil moisture and atmospheric conditions. They found that for average July soil moisture conditions, drought inception time for North Africa, the Middle East and a large part of western North America was less than 10 days.

Correlation of precipitation with antecedent soil moisture derived from a ground water balance model was computed by Fennessy and Sud (1983) for a 40 year period over the United States. Soil moisture was computed using Thornthwaite's formulation for evaporation relation as modified by Nappo (1975):

$$\frac{\partial w}{\partial t} = P - (1 - e^{-\alpha w/w_s})E_p$$

where P is precipitation, w and w_s are actual and maximum available soil water content, E_p is potential evapotranspiration. The correlation between soil moisture and subsequent precipitation appeared strongest in the western Great Plains in late summer. The high correlation in this drought prone region suggested that a feedback mechanism may be partially responsible for the maintenance of drought.

2.3 Measuring soil moisture

Given its apparent importance, soil moisture has been infrequently and incompletely measured. An exception to this situation can be found in the Soviet Union, where soil moisture has been routinely measured at hundreds of stations in agricultural areas since the 1930s, and over natural surfaces since 1967 (Vinnikov and Yeserkepova, 1991). The method currently used involves removing core samples to a depth of 1.0 to 1.5 m in 10cm segments. Each segment is weighed before and after drying to determine the mass of water contained.

Where widespread direct measurements are not available, soil moisture is often computed from a water-balance relationship. Rasmusson (1968) computed a hydrological budget for North America which included both atmospheric and land branches of the water cycle. He examined North American surface and radiosonde data from 1 May 1958 to 30 April 1963. Surface and subsurface storage change were computed as:

$$\langle\overline{\frac{\partial S}{\partial t}}\rangle = \langle\overline{P - ET}\rangle - \langle\overline{RO}\rangle$$

with evapotranspiration estimated by the methods of Thornthwaite (1964) or Budyko (1963). In the above equation angled brackets indicate spatial average, and overbars indicate time average. Using the relation for change in precipitable water:

$$-\frac{\partial W}{\partial t} = \nabla \cdot Q + (P - ET)$$

where $\nabla \cdot \bar{Q}$ is the water vapor flux divergence, change in soil moisture can also be computed without estimating evaporation using the following balance:

$$\langle\frac{\overline{\partial S}}{\partial t}\rangle = \langle\nabla\cdot\overline{Q}\rangle + \langle\frac{\overline{\partial W}}{\partial t}\rangle - \langle\overline{RO}\rangle$$

where moisture flux and change in precipitable water can be computed from radiosonde observations, and runoff computed from river flow. This vapor balance method gives an annual oscillation in soil moisture two to three times smaller than methods which rely on evaporation estimates. Rasmusson concluded that computed soil moisture is very sensitive to the treatment of evapotranspiration, but that over large areas, estimates can be considered good.

There is promise that ground moisture may also be measurable from space. Wetzel et al. (1984) attempted to deduce soil moisture by satellite using GOES infrared data. They had some success inferring soil moisture from mid-morning rate of surface temperature change with respect to absorbed solar radiation. The method was not applicable when cloud cover was present, and worked best in dry or marginal agricultural regimes. Microwave brightness temperatures, as measured by satellite, can be interpreted to yield soil moisture information in areas of sparse vegetation with greater reliability (Sellers et al., 1990). Goward (1989) has shown that surface reflectance in the visible/near infrared range can be used to infer surface soil wetness for some soil types. Again, this method is only effective where vegetation cover is scant.

3. Modeling Studies

The connection between soil moisture and the atmosphere is through evaporation. It is evaporation which is actually important to circulation. But the connection between soil moisture and evaporation is not completely understood, and often not well simulated. Soil properties, vegetation, and atmospheric conditions all affect evaporation and need to be considered for any complete simulation. Yet current state-of-the-art models rely on parameterizations or simple algorithms to represent this complex process.

Manabe et al. (1965) was the first to incorporate surface hydrology into a general circulation model (GCM), but the land and ocean surfaces were treated as completely wet with no heat capacity. Later, Manabe (1969) modified the model to predict soil moisture and snow cover.

This allowed the land surface to provide actual feedback to the atmosphere by responding to precipitation and surface heating, and altering fluxes to the atmosphere accordingly.

For many years, the state-of-the-art treatment of soil moisture was the "bucket" model of Holloway and Manabe (1971), where the land surface is treated as a grid of reservoirs which are filled with precipitation, emptied by evaporation, and may overflow to produce runoff. Soil moisture availability drops as the surface dries, making evaporation of the remaining moisture — increasingly difficult. This availability was derived by Miyakoda et al. (1979), and is incorporated with varying structures in numerous studies (see Mintz, 1984 for a representative review).

Attempts to improve upon the bucket model have taken several forms. Variations which include two or more soil layers have been developed (Deardorff, 1977; Hansen et al., 1983), and marked improvements to the implementation of the scheme have been realized (Milly, 1992). Meanwhile, some researchers have developed complex surface models which include realistic representation and distribution of vegetation, and its effects on fluxes of heat, moisture and momentum (Rind, 1984; Dickinson et al., 1986; Sellers et al., 1986). Others have pursued statistical approaches which fall between the two in complexity (Entekhabi and Eagleson, 1989; Noilhan and Planton, 1989).

3.1 Sensitivity studies

There have been numerous sensitivity studies of soil moisture using numerical models. In the interest of brevity, only a few of these studies will be reviewed here.

3.1.1 Computing soil moisture

The actual treatment of soil moisture in a GCM can take many forms, and the various forms can be compared. Meehl (1984) examined the effects of specified constant soil moisture versus computed soil moisture. Computed soil moisture led to more realistic seasonal variation in precipitation, particularly in the tropics. Hunt (1985) compared several interactive methods of parameterizing soil hydrology: the simple bucket model of Holloway and Manabe (1971), the two-layer formulation of Hansen et al. (1983), and the two-layer method of Deardorff (1977) which has an extremely shallow (5 mm) upper layer. Deardorff's formulation was found to give the most realistic results. Meehl and Washington (1988) compared the soil moisture sensitivity of two different GCMs with the same bucket soil representation. It was found that the soil moisture climatologies of the National Center for Atmospheric Research (NCAR) model and the model of the Geophysical Fluid Dynamics Laboratory (GFDL) were quite different, due to the differences between the models in the treatment of radiation and other surface properties.

3.1.2 Simple models

Walker & Rowntree (1977) used a tropical channel model to examine soil moisture effects in sub-Saharan Africa. The channel spanned 36°N to 16°S, and ran from 0°E to 32°E with cyclic boundary conditions on the east and west ends. The soil was represented by a 15 cm bucket model. The simplified surface consisted of land north of 6°N, ocean south. In Case 1 the soil moisture was initially set to 0 cm in the Sahara (14°N-32°N), and 10 cm elsewhere. In Case 2 soil moisture was initially 10 cm everywhere. In case 1, the dry soil region remained dry; in the Sahel region (6°-14°N), evaporation exceeded precipitation by 1.4 mm d^{-1} — a drought situation. In Case 2 the Sahara region gradually dried out, but precipitation spread into the region from the Sahel, which became wetter. After about day 12 precipitation equalled evaporation. The two cases reached different quasi-steady states which reflected the persistence of initial soil moisture. This highly simplified system became intransitive.

Gutman (1984) used a zonally averaged steady-state hemispheric annual-mean model to look at fixed and computed moisture, which he said was analogous to running without and with biofeedback respectively. In specific latitude bands boundary conditions were chosen to simulate desertification, deforestation, and irrigation. The boundary anomalies were held constant in the specified regions in both fixed and biofeedback runs. He found that desertification and deforestation experiments gave similar responses: reduced absorbed radiation, evapotranspiration and precipitation with a concomitant increase in adjacent areas. Irrigation had the opposite effect. Biofeedback produced changes in latitude belts adjacent to anomalies which were of the same order of magnitude as the changes produced by the anomalies themselves. Gutman concluded that biofeedback does not change sign of precipitation response in regions adjacent to anomalies, but can either amplify or moderate the change. In this simple model, perturbations do not modify climate enough to allow continued persistence. A non-stationary model would be needed to study evolution of changes.

3.1.3 GCMs with global anomalies

Shukla and Mintz (1982) performed two summer integrations with extreme soil moisture conditions - one with perpetually saturated ground and one with perpetually dry ground. Significant differences in the global patterns of surface pressure, surface temperature and precipitation were found. As shown in Figure 2, precipitation was greatly reduced and surface temperatures increased as much as 30°C in the dry soil case as compared to the wet soil case. The only region where precipitation was enhanced was over the monsoon region of southern Asia.

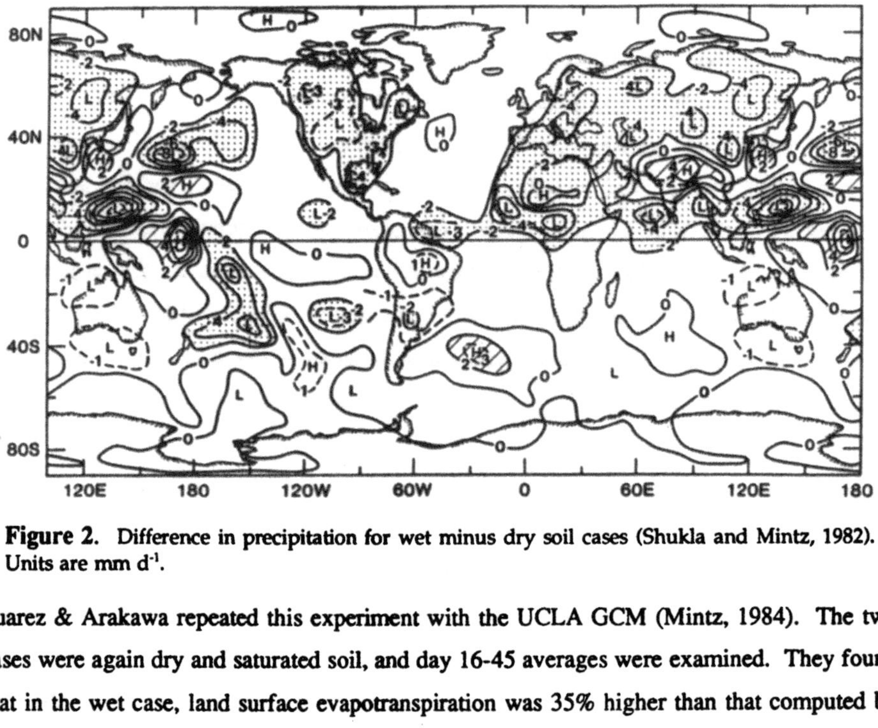

Figure 2. Difference in precipitation for wet minus dry soil cases (Shukla and Mintz, 1982). Units are mm d^{-1}.

Suarez & Arakawa repeated this experiment with the UCLA GCM (Mintz, 1984). The two cases were again dry and saturated soil, and day 16-45 averages were examined. They found that in the wet case, land surface evapotranspiration was 35% higher than that computed by Shukla and Mintz. Precipitation was nearly equal to evaporation over land. In the dry case, almost no rain fell over the continents, except over central Africa. Moisture convergence existed over some land areas but did not produce rain. Mintz (1984) conjectured that discrepancies with Shukla and Mintz were due to model differences, especially in the parameterizations of the planetary boundary layer and clouds. The UCLA model had fewer clouds, so surface radiation was stronger and evaporation was larger. The lack of precipitation in areas of moisture convergence was apparently due to some of the moisture in the PBL over land being transferred to the free atmosphere by mixing from an unusually strong diurnal cycle over dry land in the UCLA model.

Carson and Sangster (1981) performed GCM experiments with globally saturated and dry initial soil moisture conditions in a bucket model. Evidence of the initial anomalies was still visible in the day 21-50 average precipitation, and some areas still reflected the initial anomalies after 200 days.

3.1.4 GCMs with regional anomalies

There have been numerous experiments with regional scale anomalies. Yeh et al. (1984), using idealized land-sea distributions, found that wet soil moisture anomalies enhanced precipitation only in the mid-latitudes (which were already rainy). However, the anomalies were least persistent in the tropics. This first result has also been found for anomalies over Europe (Rowntree and Bolton 1978; 1983). Simulations with North American soil moisture anomalies (Rind, 1992; Oglesby and Erickson, 1989) imply that droughts can be intensified or prolonged by locally low soil moisture, especially in the interior of the continent. Sud and Smith (1985) found that reduced soil moisture over India seemed to have no effect on precipitation. Experiments in other subtropical areas give mixed results (Sud et al., 1982; Sud and Fennessy, 1984; Kitoh et al., 1988) implying that regional circulation patterns may overwhelm the forcing of the atmosphere by soil moisture anomalies in some areas.

3.1.5 Persistence of anomalies

Delworth and Manabe (1988) used a low-resolution GCM to investigate the character of the persistence of soil moisture anomalies. The GCM used a 15 cm bucket model of hydrology and integrations were carried out for 50 years. Monthly averages from the integration were subtracted from the global fields of soil moisture and precipitation so that only a 50 year record of anomalies remained. It was found that the spectra of precipitation anomalies were nearly white at all latitudes, while the soil moisture spectra were red, with most of the power at very long periods. The redness of the spectra increased from equator to pole. Thus, the relationship between soil moisture and precipitation is very closely approximated by a first-order Markov process with precipitation as the forcing, and potential evaporation providing the damping. For their model, half of the variance in soil moisture was at periods greater than 7.5 months in the tropics and subtropics (3°S-31°N), over 12 months in the mid-latitudes (31°-54°N), and 20 months at high latitudes (54°-76°N). However, the model lacked a diurnal cycle, and some important sources of feedback which may alter the persistence of soil moisture anomalies, such as interactive cloudiness and the seasonal variation of potential transpiration, which is caused by the annual cycle of vegetation.

3.1.6 Biosphere models

Sensitivity studies, which have attempted to measure GCM response to the inclusion of vegetation parameterizations, have given tangible if not systematic results. Sato et al. (1989) found that coupling of the Simple Biosphere (SiB) model to a GCM corrected evaporation errors of bucket hydrology. Henderson-Sellers et al. (1990) found that their Biosphere Atmosphere Transfer Scheme (BATS) also seemed to reduce some of the gross errors of the bucket model. In tests of the sensitivity of BATS to various soil characteristics (Wilson et al., 1987), sensitivity to soil texture and upper soil layer depth was found to be high. Although biosphere models depict the process of evapotranspiration more realistically than simpler schemes, they do not necessarily improve on simple schemes in all situations. Also, the vegetation parameterizations in current biosphere models do not react to anomalies or trends in climate, and cannot simulate interannual variability in vegetation cover or vigor.

3.2 Prediction

The overwhelming majority of modeling studies have focused on the sensitivity of the atmospheric response to changes in soil moisture or its formulation. However, a few have examined the feasibility of soil moisture as a predictor of short-term climate. Rind (1982) has investigated the predictive capability of spring soil moisture with relation to summer temperatures and precipitation over the United States. In particular, low springtime soil moisture can be looked upon as a precursor to a hot dry summer. GCM integrations show that while precipitation is reduced, evaporation is reduced more. Thus, "E minus P" decreases and the soil may recharge, erasing the dry anomaly. This processes limits predictability to two or three months - more where the prevailing circulation does not advect moisture from an oceanic source.

The mechanism by which soil moisture anomalies may aid development and persistence of drought in a GCM was explored by Oglesby and Erickson (1989). They determined that a reduction in soil moisture leads to increased surface temperature. The lower atmosphere is heated and ridging occurs aloft. Low level moisture advection is a controlling factor in maintenance of the drought. The degree to which the model circulation is in equilibrium with the soil moisture anomaly determines how quickly the anomaly is diminished.

Meehl (1984) found that inclusion of predicted soil moisture increased the accuracy of monsoon simulations in the NCAR Community Climate Model. He attributed this to the role of soil moisture as a positive feedback over inland regions. This result has been confirmed by Fennessy (personal communication) in experiments where initial soil wetness over India is set very high. Rainfall increased over India as compared to a control run. An experiment with reduced initial soil moisture showed less of a change, but also resulted in increased rainfall. In this case, increased surface heating may be increasing convergence over India. This change in circulation concentrates more moisture over India, similar to what was observed in the study of Shukla and Mintz (1982) with zero soil moisture.

Fennessy and Shukla (1992) performed a similar experiment in a GCM with interactive soil moisture and biosphere. Initial soil moisture was set globally to either climatological values or proxy observed soil wetness derived from the analysis-forecast system of the European Centre for Medium-Range Weather Forecasting (ECMWF). An ensemble of seasonal model integrations were initialized from observed atmospheric states on each of the first three days of June in both 1987 (a non-drought year in central North America) and 1988 (a severe drought year). Soil moisture anomalies were highly persistent. Figure 3 shows the 1 June 1988 initial soil wetness anomalies (ECMWF minus climatology), and the seasonal mean (JJA) soil wetness differences between the two simulations. The integrations with ECMWF initial soil moisture produced a reasonable simulation of 1988 North American drought in both precipitation and surface temperature anomalies, as compared to integrations with climatological initial soil moisture. The simulation of 1988 versus 1987 interannual variability is also ameliorated by use of the "observed" initial soil wetness.

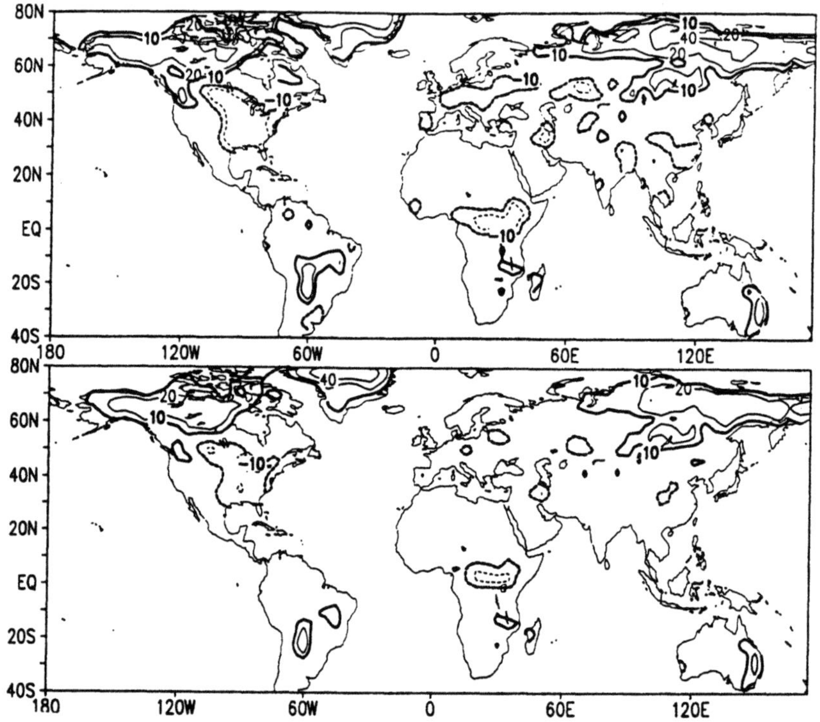

Figure 3: 1 June 1988 soil wetness anomalies (top) and simulated JJA mean soil wetness differences (bottom). Contours are -20, -10, 10, 20, and 40 percent.

In order to use soil moisture as a predictor or indicator of future climate anomalies, accuratesoil moisture measurements must be available to initialize the models, and the models must accurately predict soil moisture. The problems of obtaining comprehensive soil moisture measurements was discussed earlier. Yang et al. (1992) found that relatively small errors in initial soil moisture specification can contribute to sizeable short-term errors in surface air temperature and relative humidity. This is because most of the change in maximum diurnal surface temperature as a function of soil moisture occurs across a narrow range of wetnesses. Yang et al. (1991) have developed a method to correct initial soil moisture based on the computed error of the corresponding surface temperature as predicted by a GCM. Figure 4 shows the mean surface air temperature for the first five days of an ensemble of three summertime forecasts initialized with uncorrected soil moistures. Errors as great as 5°C occur in the semi-arid regions of the western Great Plains. When initial soil moistures are corrected, errors are reduced significantly (Figure 5). Root-mean-square errors also are reduced.

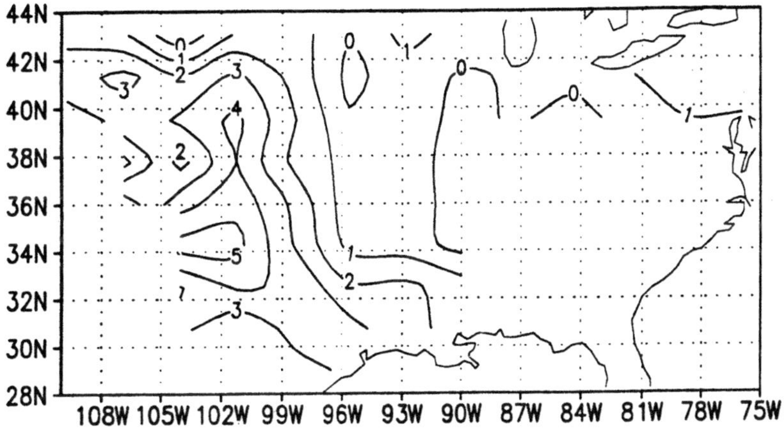

Figure 4: Error in mean surface air temperature for the first five days of forecast with uncorrected soil wetness (Yang et al., 1992). The contour interval is 1°C.

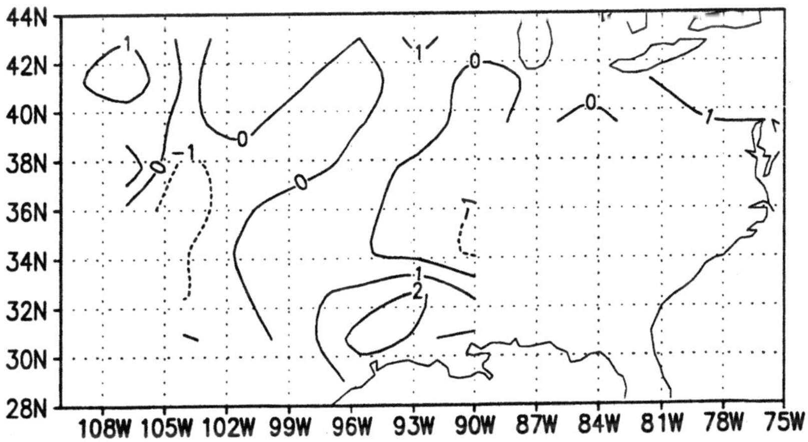

Figure 5: As in Figure 4, except for corrected soil moisture.

Verification of model performance in simulating seasonal soil moisture variations has been conducted by Vinnikov and Yeserkepova (1991). GFDL, Oregon State University (OSU) and United Kingdom Meteorological Office (UKMO) model results were compared to observations over the Soviet Union. All three models use bucket parameterizations, and all grossly underpredicted soil moisture for summer.

3.3 Subgrid variability

Recently there has been a great deal of interest in the role of subgrid scale variability of surface moisture and vegetation in modeling. A number of ways of addressing the problem of representing small-scale variety in large grid boxes have been developed. In the SiB surface model, Dorman and Sellers (1989) use plurality to determine the vegetation and soil type for an entire grid box. Abramopoulos et al. (1988) use area-weighted means of soil and vegetation parameters in a somewhat simpler surface representation. A gaussian distribution of subgrid variations of soil moisture and evaporation is used by Wetzel and Chang (1988) to represent the spottiness of precipitation in wetting a large grid box. Entekhabi and Eagleson (1989) use an exponential distribution of precipitation and soil moisture. Koster and Suarez (1991) have developed a mosaic treatment by which fluxes are computed for each vegetation and soil type found within a given grid box using identical upper boundary conditions. The results from each "tile" of the mosaic are then area-weighted and averaged together to supply a single set of boundary flux values to the atmospheric model.

4. Summary and Conclusions

Soil moisture directly affects the partitioning of energy at the surface between latent and sensible heating. Where soil moisture is high, evaporation will predominate, adding to atmospheric moisture content. Where it is low, the land surface will warm under the influence of radiational heating. Thus, it is not soil moisture which directly affects the atmosphere, but latent and sensible heating which is modulated by soil moisture.

Soil moisture is difficult to quantify in terms of its affect on the atmosphere. Many other factors, such as vegetation, soil characteristics and ambient conditions alter the transfer of moisture from soil to the air. Also, soil moisture is very difficult to measure directly due to its heterogeneity at small scales. There is observational evidence that soil moisture correlates to future rainfall and temperature in certain instances. These correlations seem to be valid only out to three months, and are highly dependent on season and location.

Modeling studies show that strong perturbations in soil moisture on global or regional scales can affect atmospheric circulation, and persist for several months. However, the role of soil moisture in generating long term climate variability is not well understood. There has been no observational evidence that soil moisture anomalies in an otherwise unaltered surface can persist for time scales beyond a year, nor can they affect interannual climate. Yet changes in surface vegetation such as deforestation (Dickinson and Henderson-Sellers, 1988; Lean and Warrilow, 1989; Nobre et al., 1991) and desertification (Xue and Shukla, 1992) can change the climatological values of soil moisture, as well as surface roughness and albedo. These changes may then alter climate in significant ways. This may be the most important and ominous manifestation of soil moisture repercussions on climate.

REFERENCES

Abramopoulos F, Rosenzweig C and Choudhury B (1988) Improved ground hydrology calculations for global climate models (GCMs): Soil water movement and evapotranspiration. *J Climate* 1:921-941

Anthes R (1984) Enhancement of convective precipitation by mesoscale variations in vegetative covering in semiarid regions. *J Climate Appl Meteor* 23:541-554

Barnston AG and Schickedanz PT (1984) The effect of irrigation on warm season precipitation in the southern Great Plains. *J Climate Appl Meteor* 23:865-888

Budyko MI (1963) *Atlas Teplovogo Balansa Zemnogo Shara*, U.S.S.R. Glavnaya Geofizicheskaya Observatorya, Moscow, 69pp

Carson DJ and AB Sangster (1981) The influence of land-surface albedo and soil moisture content on general circulation model simulations. *Research Activities in Atmospheric and Ocean Modelling*, I D Rutherford, ed. GARP/WCRP Report No. 2, 5.14-5.21

Deardorff JW (1977) A parameterization of ground-surface moisture content for use in atmospheric prediction models. *J Appl Met* 16:1182-1185

Dickinson RE, Henderson-Sellers A, Kennedy PJ and Wilson MF (1986) Biosphere-atmosphere transfer scheme (BATS) for the NCAR Community Climate Model. National Center for Atmospheric Research, Boulder, CO. Tech Note/TN275+STR

Dickinson RE and Henderson-Sellers A (1988) Modelling tropical deforestation: A study of GCM land-surface parametrizations. *Quart J Roy Meteor Soc* 114:439-462

Dickinson RE and Kennedy PJ (1991) Climate effects of Amazon deforestation revisited. *Proc. of the AGU-MSA 1991 Spring Meeting*, Suppliment to Trans Amer Geophys Union American Geophysical Union Washington 82

Dorman JL and Sellers PJ (1989) A global climatology of albedo, roughness length and stomatal resistance for atmospheric general circulation models as represented by the simple biosphere model (SiB). *J Appl Meteor* 28:834-855

Eagleson PS (1991) Hydrologic science: A distinct geoscience. *Rev Geophys* 29:237-248

Entekhabi D and Eagleson PS (1989) Land surface hydrology parameterization for atmospheric general circulation models including subgrid scale spatial variability. *J Climate* 2:816-831

Fennessy MJ and Sud YC (1983) A Study of the Influence of Soil Moisture on Future Precipitation. *NASA Tech. Memo 85042* Goddard Space Flight Center, Greenbelt MD 123p

Fennessy MJ and Shukla J (1992) Influence of initial soil wetness on prediction of global atmospheric circulation and rainfall. J Climate (submitted)

Goward SN (1989) Satellite Bioclimatology. *J Climate* 2:710-720

Gutman G (1984) Numerical experiments on land surface alterations with a zonal model allowing for interaction between the geobotanic state and climate. *J Atmos Res* 41:2679-2685

Hansen J, Russell G, Rind D, Stone S, Lacis A, Lebedeff S, Ruedy R and Travis L (1983) Efficient three-dimensional global models for climate studies: Models I and II. *Mon Wea Rev* 111:609-662

Henderson-Sellers A, Pitman AJ and Dickinson RE (1990) Sensitivity of the surface hydrology to the complexity of the land-surface parameterization scheme employed. *Atmósfera* 3:183-201

Holloway JL and Manabe S (1971) Simulation of climate by a global general circulation model. I. Hydrologic cycle and heat balance. *Mon Wea Rev* 99:335-370

Hunt BG (1985) A modeling study of some aspects of soil hydrology relevant to climatic modelling. *Quart. J Roy Meteor Soc* 111:1071-1085

Karl TR (1983) Some spatial characteristics of drought duration in the United States. *J Climate Appl Meteor* **22**:1356-1366

Karl TR (1986) The relationship of soil moisture parameterizations to subsequent seasonal and monthly mean temperature in the United States. *Mon Wea Rev* 114:675-686

Kitoh A, Yamakazi K and Tokioka T (1988) Influence of soil moisture and surface albedo changes over the African tropical rain forest on summer climate investigated with the MRI-GCM-I. *J Meteor Soc Japan* **66** 65-86

Koster RD and Suarez MJ (1992) Modeling the land surface boundary in climate models as a composite of independent vegetation stands. *J Geophys Res* **97**:2697-2716

Lean J and Warrilow DA (1989) Simulation of the regional climatic impact of Amazon deforestation. *Nature* **342**:411-413

Manabe S (1969) Climate and the circulation, I. The atmospheric circulation and the hydrology of the earth's surface. *Mon Wea Rev* **97**:739-774

Manabe S, Smagorinsky J and Strickler RF (1965) Simulated climatology of a general circulation model with a hydrological cycle. *Mon Wea Rev* **93**:769-798

Meehl GA (1984) Soil moisture, a simple mixed layer ocean and the southern hemisphere semiannual oscillation in the NCAR Community Climate Model. *Studies in Climate,* H. van Loon, ed., NCAR/TN-227+STR, National Center for Atmospheric Research, NTIS No. PB84 196385:115-150

Meehl GA and Washington WM (1988) A comparison of soil-moisture sensitivity in two global climate models. *J Atmos Sci* **45**:1476-1492

Milly PCD (1992) Potential evapotranspiration and soil moisture in general circulation models. *J Climate* **5**:209-226

Mintz Y (1984) The sensitivity of numerically simulated climates to land-surface boundary conditions. *The Global Climate,* J. Houghton, ed., Cambridge University Press, 79-105

Miyakoda K, Hembree GD and Strickler RF (1979) Cumulative results of extended forecast experiments. II: Model performance for summer cases. *Mon Wea Rev* **107**:395-420

Namias J (1959) Persistence of mid-troposheric circulations between adjacent months and seasons. *The Atmosphere and the Sea in Motion,* B. Bolin, ed., Rossby Memorial Volume 240-248

Namias J (1960) Factors in the initiation, perpetuation and termination of drought. *[Extract of publication No. 51,* I.A.S.H. Commision of Surface Waters 81-94

Namias J (1989) Cold waters and hot summers. *Nature* **338**:15-16

Nappo CJ (1975) Parameter of surface moisture and evaporation rate in a planetary boundary layer model. J Appl Meteor **14**:289-296

Nobre CA, Sellers PJ and Shukla J (1991) Amazon deforestation and regional climate change. *J Climate* **4**:957-988

Noilhan J and Planton S (1989) A simple parameterization of land surface processes for meteorological models. *Mon Wea Rev* **117**:536-549

Oglesby RJ and Erickson III DJ (1989) Soil moisture and persistence of North American drought. *J Climate* **1**:1362-1380

Palmer WC (1965) Meteorological drought. Res Pap 45 U.S. Weather Bureau, 58pp

Rasmusson EM (1968) Atmospheric water vapor transport and the water balance of North America II. Large-scale water balance investigations. *Mon Wea Rev* **96**:720-734

Rind D (1982) The influence of ground moisture conditions in North America on summer climate as modeled in the GISS GCM. *Mon Wea Rev* **110**:1487-1494

Rind D (1984) The influence of vegetation on the hydrologic cycle in a global climate model. *Climate Processes and Climate Sensitivity,* J. E. Hansen and T. Takahashi, eds., Geophysical Monograph 26 Maurice Ewing 5:73-91

Rowntree PR and Bolton JA (1978) Experiments with soil moisture anomalies over Europe. *The GARP Programme on Numerical Experimentation:* Research Activities in Atmospheric and Ocean Modelling, R. Asselin ed., Report No. 18 WMO/ICSU Geneva 63

Rowntree PR and Bolton JA (1983) Effects of soil moisture anomalies over Europe in summer. *Variations in the Global Water Budget,* A. Street-Perrott et al., eds, D Reidel 447-462

Salati E and Nobre CA (1991) Possible climatic impacts of tropical deforestation. *Climatic Change* (in press)

Sato N, Sellers PJ Randall DA Schneider EK Shukla J Kinter III JL Hou Y-T and Albertazzi E (1989) Effects of implimenting the Simple Biosphere model in a general circulation model. *J Atmos Sci* **46**:2757-2782

Sellers PJ, Mintz Y Sud YC and Dalcher A (1986) A simple biosphere model (SiB) for use within general circulation models. *J Atmos Sci* **43**:505-531

Sellers PJ, Rasool SI and Bolle H-J (1990) A review of satellite data algorithms for studies of the land surface. *Bull Amer Meteor Soc* **71**:1429-1447

Serafini YG and Sud YC (1982) A model for the inception time of an agricultural drought. *Research Activities in Atmospheric and Ocean Modelling,* ID Rutherford, ed. GARP/WCRP Report No 3:5.38-5.43

Shukla J and Mintz Y (1982) Influence of land-surface evapotranspiration on the earth's climate. *Science* 215:1498-1501

Sud YC Fennessy M and Shukla J (1982) Influence of evapotranspiration anomalies in semi-arid regions on July circulation. *Research Activities in Atmospheric and Ocean Modelling.* ID Rutherford, ed. GARP/WCRP Report No 3:5.30-5.33

Sud YC and Fennessey MJ (1984) Influence of evaporation in semi-arid regions on the July circulation: A numerical study. *J Climatology* 4:383-398

Sud YC and Smith WE (1985) Influence of local land-surface processes on the Indian monsoon: A numerical study. *J Climate Appl Meteor* **24**:1015-1036

Thornthwaite CW and Associates (1964) Average climatic water balance data of the continents: Part VII. United States. *Publications in Climatology*, Laboratory of Climatology, Centerton, NJ **17**:419-615

Vinnikov K Ya and Yeserkepova IB (1991) Soil moisture, empirical data and model results. *J Climate* **4**:66-79

Walker J and Rowntree PR (1977) The effect of soil moisture on circulation and rainfall in a tropical model. *Quart J Roy Meteor Soc* **103**:29-46

Walsh JW, Jasperson H and Ross B (1985) Influences of snow cover and soil moisture on monthly air temperature. *Mon Wea Rev* **113**:756-768

Wetzel PJ and Chang J-T (1988) Evapotranspiration from nonuniform surfaces: A first approach for short-term numerical weather prediction. *Mon Wea Rev* **116**:600-621

Wetzel PJ, Atlas D and Woodward RH (1984) Determining soil moisture from geosynchronous satellite infrared data: A feasibility study. *J Climate Appl Meteor* **23**:375-391

Wilson MF, Henderson-Sellers A, Dickinson RE and Kennedy PJ (1987) Sensitivity of the Biosphere-Atmosphere Transfer Scheme (BATS) to the inclusion of variable soil characteristics. *J Climate Appl Meteor* **26**:341-362

Xue Y and Shukla J (1992) A numerical experiment to study the influence of changes in the land surface properties on Sahel climate: Part 1 - Deforestation. *J Climate* (submitted)

Yang R, Fennessy MJ and Shukla J (1992) The influence of the variation of initial soil wetness on the surface weather and atmospheric circulation. *Mon Wea Rev* (submitted)

Yeh T-C, Wetherald RT and Manabe S (1984) The effect of soil moisture on the short-term climate and hydrology change - A numerical experiment. *Mon Wea Rev* **112**:474-490

OBSERVATIONAL AND MODELLING STUDIES OF THE INFLUENCE OF LAND SURFACE ANOMALIES ON ATMOSPHERIC CIRCULATION (FUTURE DIRECTIONS)

L. Dümenil and L. Bengtsson
Max-Planck-Institut für Meteorologie
Bundesstraße 55, 2000 Hamburg 13, F.R.G.

1. INTRODUCTION

Over the years, the numerical models used for climate studies have evolved from one-dimensional energy balance models to coupled three-dimensional models, comprising the various sub-systems of the climate systems; atmosphere, ocean, sea-ice, glaciers, lakes on continents, the land surfaces including their biomass (in order to represent the full carbon cycle), etc.

The questions that will have to be addressed by numerical modelling are not restricted to the effects of increasing greenhouse gases in the atmosphere but cover an exceedingly large range of questions associated with global change (land use change, degradation of soils, desertification, deforestation, etc.). In the context of climate change the question is posed what impact with regard to water resources and agriculture the global change scenarios will have on biomes and human life. A primary requirement is a further reduction of systematic model errors and the development of more realistic models having a higher degree of complexity in order to study the variety of questions that are being posed today. The representation of land surface processes is one of the areas where improvement is required.

NATO ASI Series, Vol. I 6
Prediction of Interannual Climate Variations
Edited by J. Shukla

2. THE IMPORTANCE OF LAND SURFACE PROCESSES

Although land surface processes have for a long time only been considered as providing boundary conditions for the atmosphere models, it has been recognized that land surfaces may play an important role in the determination of climate and climate variability on at least the regional scale.

The skill of the simulation of present-day climate is to a considerable degree dependent on the correct specification of the boundary conditions. Although the basic dynamics of atmosphere climate models are not extremely sensitive to errors in the boundary conditions in short-term integrations, the errors may increase to an unacceptable level when the integrations are extended over longer periods. The various components of the hydrological cycle e.g. depend on the model's large-scale dynamics as well as on the surface boundary conditions by which the dynamics strongly interact. Errors associated with the representation of the hydrological cycle are connected to the distribution of energy in the model and therefore have a potentially strong impact on the general circulation. Experimentation is under way to investigate which climate parameters are affected by the boundary conditions during a long-term simulation. Examples from a preliminary experiment where a soil moisture of 100% was prescribed everywhere on the continents under permanent July conditions show that the global zonal mean temperature error increases only in certain regions in the lower troposphere. A more immediate effect is seen on the quantities that are directly connected to the hydrological cycle. Fig. 1 illustrates this for the precipitation field. While the vertical temperature distribution in the model is hardly affected even after such a long time, the model precipitation field is completely different from the observed. This supports our view that land surface conditions are highly important in longterm integrations and that they have to be represented to a certain degree of accuracy in order to obtain a faithful simulation of present-day climate.

The correct representation of land surface processes is tied to the solution of three fundamental problems:

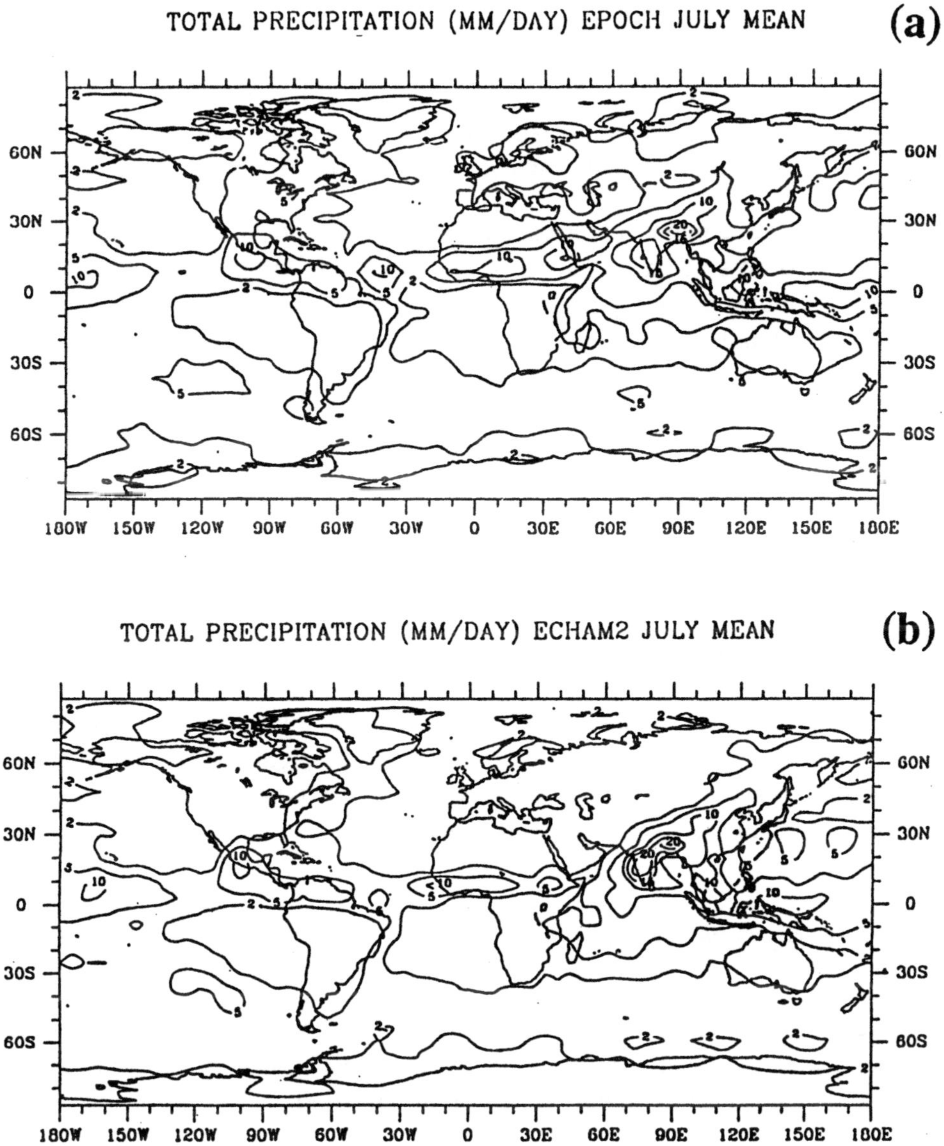

FIGURE 1: Global distribution of precipitation (mm/d) for an integration comprising 360 diurnal cycles under perpetual July insolation conditions and July climatological sea surface temperatures for a) the control integration with interactive soil moisture (top) and b) the integration where soil moisture over land surfaces is permanently at 100% saturation.

(i) Methods must be found to incorporate for the fine horizontal and time scales at which these processes may occur and to the variability of land surface characteristics. Albedo, soil types, potential evaporation etc. are difficult to specify at the required high resolution for all the areas of the globe.

(ii) The process of evapotranspiration is complex and even if the problems are treatable at the very small-scale unit of a plant or soil point, the parameterization of the processes in global models would always have to be strongly simplified.

(iii) A proper parameterization is required for the contributions by sub-grid scale organized motion (e.g. mesoscale circulation).

It is obvious that land surface processes can be expected to be simulated correctly only if the systematic errors due to other parts of the model are minimized. Errors in model dynamics, the cloud parameterization or the radiation scheme often interfere with sophisticated land surface schemes. They are in fact responsible for part of the errors in the surface energy balance that subsequently causes the surface components of the hydrological cycle and energy balance to deviate from observations. Fig. 2 is taken from a paper by Dickinson (1989) who discusses this problem.

3. PLANNED EXPERIMENTATION

In order to investigate some of the interactions between the land surfaces and the other components of the climate system, several plans for numerical experimentation have been initiated.

MPI will be involved in the European Community Programme on Climate Change and Hazards (EPOCH), which is committed to study the effects of land surface processes, in particular soil moisture effects, on the general circulation. The emphasis here will be on the European area and experiments are envisaged for present-day climate as well as CO_2 scenarios with increased greenhouse gases.

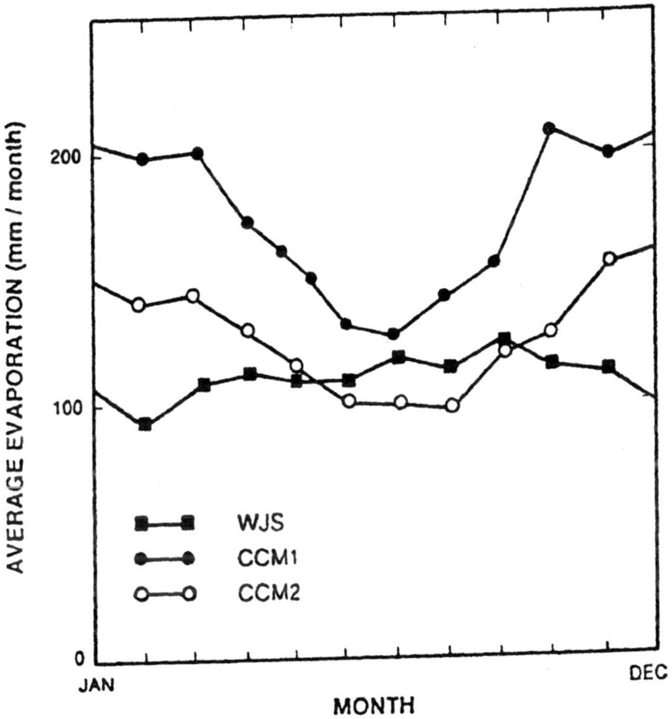

FIGURE 2: Monthly average values of total evaporation for Amazonian rain forest measured at a single point near Manaus, marked WJS, and two GCMs simulations for a co-located grid square made with the Biosphere-Atmosphere Transfer Scheme (BATS) in different versions of the NCAR community Climate Model, marked CCM1 and CCM2 (from Dickinson, 1989).

Three European GCMs will be compared in studies relating to their sensitivity to vegetation cover, values of stomatal resistance and the relative importance of land surface characteristics versus the influence of sea surface temperatures.

Studies with the early version of the Hamburg low resolution climate model have stressed the importance of land surface processes in connection with the inter-annual variability of the Asian monsoon. The ECHAM/T42 model will be used to study the relative response of the monsoon to sea surface temperatures and land surface parameters in cooperation with the TOGA Monsoon Experimentation Group.

Diagnostics of the performance of the atmosphere and coupled atmosphere/ocean model are currently under way. Particularly interesting is the question if the model is able to reproduce classification of climatic regions such as Köppen's for the present-day climate. This would require that the model successfully simulates several parameters in combination at the regional scale and also during the annual cycle. Their faithful representation will be a pre-requisite for the planned implementation of a biome model (e.g. Prentice et al., 1991) to be used interactively during the model integration and which would provide the kind of parameters that are required for climate change impact research (hydrology, agriculture, forestry, etc.).

4. THE HAMBURG CLIMATE MODEL

4.1. General

The Hamburg climate model, ECHAM2, is a state-of-the-art climate model which can be fully coupled to a global ocean model. An outline of the model and selected atmospheric climate diagnostics are described in Roeckner et al. (1991). For the experiments described below the model has been run in a T21 resolution.

With regard to land surface processes, the model has an "intermediate" level of complexity which is consistent with the low resolution and the demand for a fast vector code. The processes of evaporation, infiltration, runoff and river-runoff, snow packs on the ground and the specification of albedo are examples of land-surface processes that the ECHAM2 model considers. While the specifications of albedo and soil hydrology are more advanced than in other models, the influence of vegetation cover and soils on evaporation follows the approach taken by Blondin (1984) at ECMWF which is conceptually simpler than for example the BATS model (Dickinson, 1986) or the SiB model (Sellers et al., 1986).

The approach taken in representing the soil hydrology in the ECHAM2 model may, however, serve as an example of how the various aspects of the land surface processes should be treated in later versions of the model. We shall therefore discuss the scheme employed here in more detail.

4.2. Demands of the climate model

The specification of the soil hydrology scheme in the ECHAM2 model is a first attempt to comply with the demands of the climate model for a more detailed description of the hydrological process at the continental surfaces. Climate models require the computation of a runoff that consists of two components. One is the calculation of the local excess of precipitation over infiltration over a whole grid area. The second is a slowly acting process which stems from the drainage of soil water for values of soil moisture close to saturation. Both components of local runoff act on different time scales and contribute to the river-runoff which is organized in river catchments.

GCMs so far have mainly represented the local aspect of soil hydrology in a rather crude way by the so-called 'bucket'-model (Manabe, 1969). As an intermediate step before full large-scale hydrological models become available, we decided to use the Arno rainfall-runoff scheme (Dümenil and Todini, 1991), a simplified scheme that is based on hydrological information which can be applied everywhere on the globe. It should account for grid area heterogeneity as much as possible, because the local runoff depends crucially on the distribution of orography, vegetation, soil types, lakes and rivers within a grid cell. At the same time it must be able to represent the large-scale variation of these characteristics on the continental scale. Ideally, the scheme

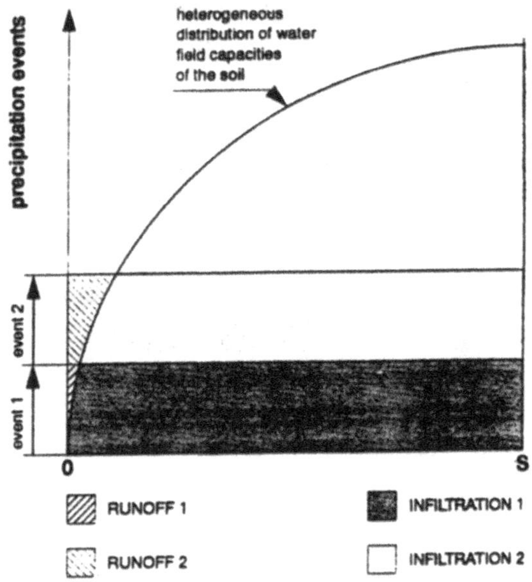

FIGURE 3: Representation of the Arno rainfall-runoff scheme. A heterogeneous distribution of local water field capacities for the grid area defines regions within a grid cell, which quickly reach saturation during a precipitation event and where runoff occurs. The remaining water goes into infiltration (Dümenil and Todini, 1991).

should not be relying on assumptions and parameters that are only valid for present-day climate, but the model should be applicable also in paleo- or changed climate scenarios. At the same time it should not rely on parameters for which there is no or insufficient global data.

Fig. 3 illustrates the basic principles of the two Arno schemes. During a precipitation event, runoff occurs in certain regions as an excess of rainfall over infiltration. The infiltration capacities or soil water field capacities depend on many factors which are heterogeneously distributed over a grid area. The Arno scheme has been developed in order to account for this heterogeneity within a catchment.

The stippled curve in Fig. 3 gives the nonlinear distribution of field capacities over the area (as a reminder, a 'bucket model' assumes a homogeneous distribution of field capacities over the whole area). Parts of the area may therefore reach saturation sooner than others. Therefore, certain parts of the area will reach saturation at once during a precipitation event. From these parts there will be runoff while the precipitation rainfall may infiltrate into the remaining grid area.

In the current model version we combine the Arno rainfall-runoff scheme with a river-runoff scheme (Sausen et al., 1991) that collects the water in large river catchments. The water is then subjected to a linear advection model which transfers it to the river mouths, where it is passed on to the control of the ocean model as freshwater input.

4.3 Validation of the hydrological cycle simulated by ECHAM2

Simulations with the current model can be easily verified with regard to their large-scale dynamical aspects and parameters for which global datasets are available. Fields such as precipitation are much more difficult to validate due to an enormously higher horizontal variability. It is therefore sensible to validate as many parameters associated with the hydrological cycle as possible in order to obtain more complementing information on the model performance. In the ECHAM2 model, the globally averaged annual hydrological cycle (Fig. 4) is found to be in good agreement with empirical estimates given by Baumgartner and Reichel (1975). The quality of the distribution of precipitation from month to month and its horizontal distribution is difficult to judge

GLOBAL ANNUAL HYDROLOGICAL CYCLE

Model: ECHAM2 (CTL) 20 - year average
Observations: Baumgartner and Reichel (1975) (OBS)

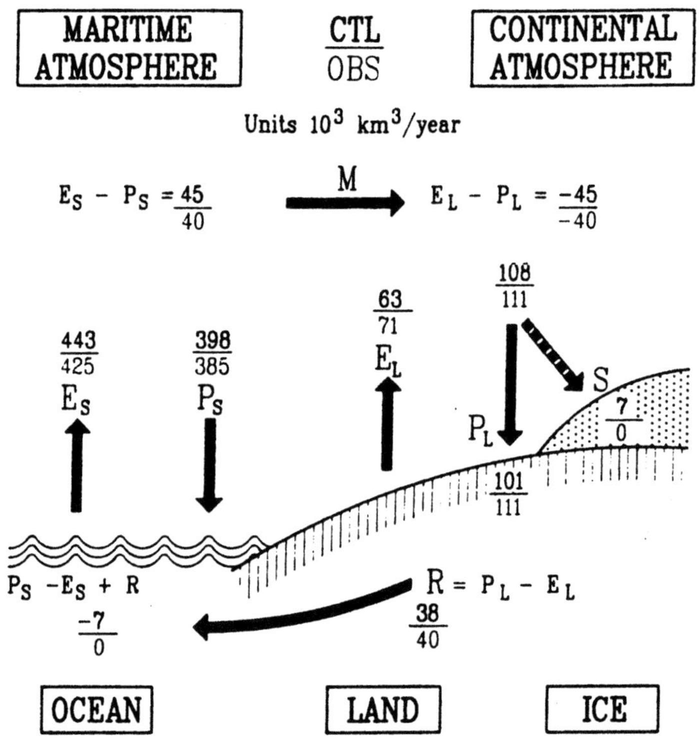

E_S, E_L Evaporation (sea, land)
P_S, P_L Precipitation (sea, land)
S Snow deposition on land ice
M Moisture transport
R Runoff

FIGURE 4: Globally averaged values of the components of the hydrological cycle. Results from the ECHAM2 (T21L19) simulations (upper figures) and climatological values (lower figures) from Baumgartner and Reichel (1975).

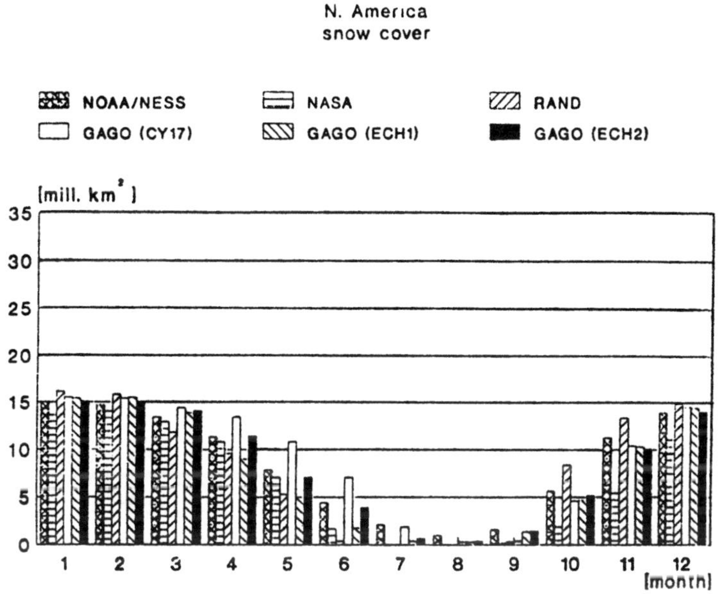

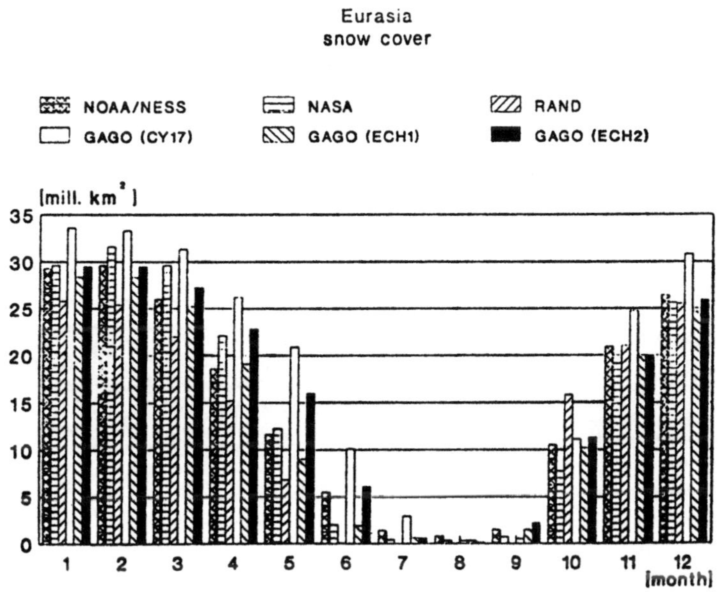

FIGURE 5a: Annual cycle of long-term averages of snow cover (in 10^6 km^2) from observations (NOAA/NESS, NASA, RAND) and GCM simulations (Cycle 17, ECHAM1 and ECHAM2 model versions) for North America and Eurasia (from Behr and Dümenil, 1991).

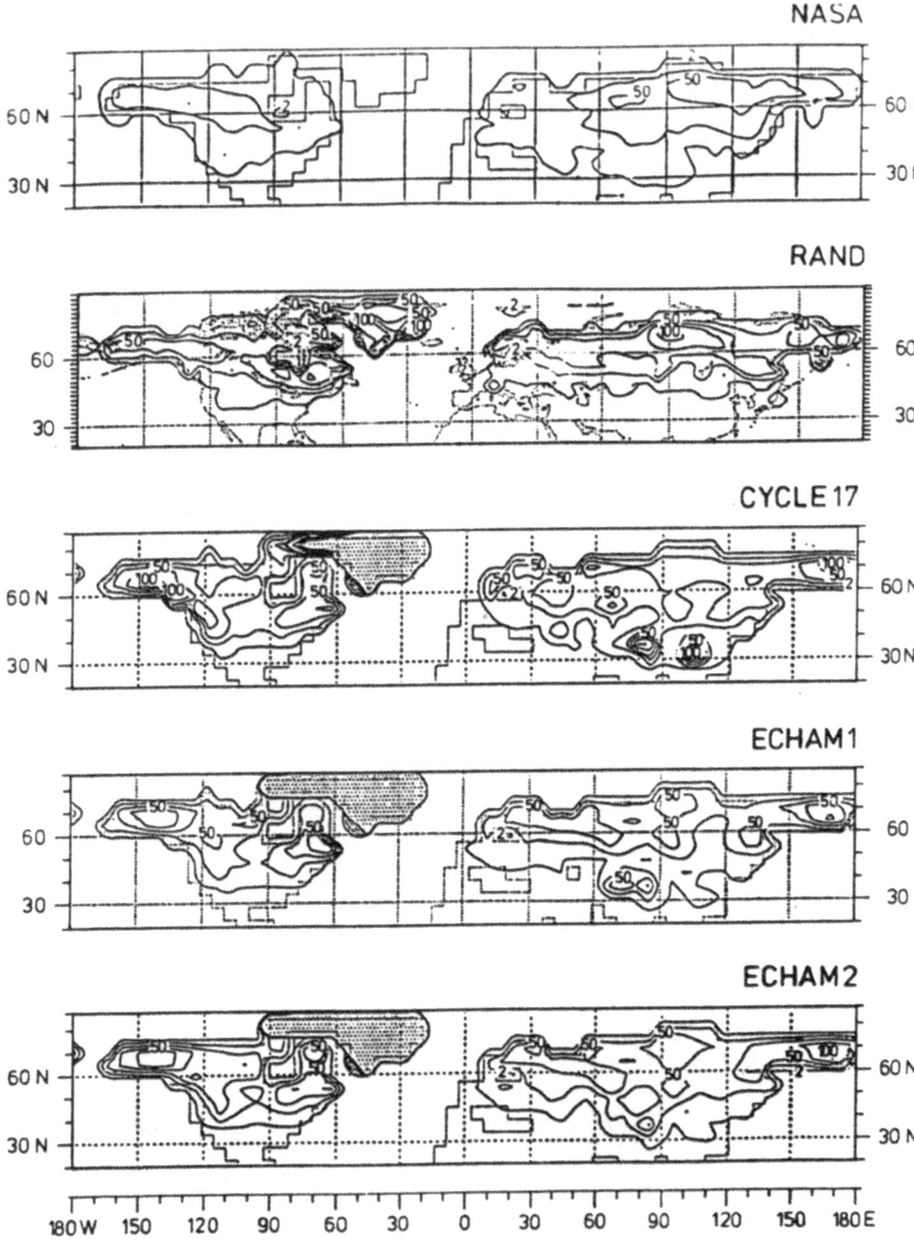

FIGURE 5b: Global distributions of long-term averages of snow depth (in cm) for the month of February for observations (NASA, RAND) and GCM simulations (Cycle 17, ECHAM1, ECHAM2) contour intervals at 2, 25, 50 ... cm (from Behr and Dümenil, 1991).

from the existing climate datasets which are of limited usefulness and leave a high degree of uncertainty. In addition we have therefore diagnosed the evolution of the model generated snow field (Fig. 5), the soil moisture field (Fig. 6) and the local rainfall-runoff (Fig. 7), and have compared the river-runoff from the model with the gauge measurements from the ten largest rivers (Fig. 8 and 9) as an independent observation.

The snow line (Fig. 5) is in good agreement with the few climatological data that exist (Behr and Dümenil, 1991). The distribution of the depth of the model snow pack is more difficult to validate. There is still a need for data on the global distribution of snow depth (water equivalent) in the annual cycle. This might be obtained from well-calibrated microwave data from remote sensing. At the same time a climatology from existing, so far inaccessible, conventional data should be compiled.

The soil moisture climatology of Mintz and Serafini (1981) computed from precipitation data is therefore not an independent dataset for validation. In snow-covered regions it does not account for example for snow melt. Apart from these areas there is good agreement between the model results and these data.

The river gauge data is presently being compiled at the Global Runoff Data Centre in Koblenz, FRG. This initiative will provide an independent data base for the validation of model output. Together with further information on the surface energy balance, these datasets may help to investigate the problem why the model snow-field disagrees with observations only during the melting period in the spring.

The comparison of model results with this dataset can only be preliminary, because a quality control of the observations has not yet been possible. Local differences may result from systematic errors in the model precipitation in adequation of the rainfall-runoff scheme and the river-runoff scheme.

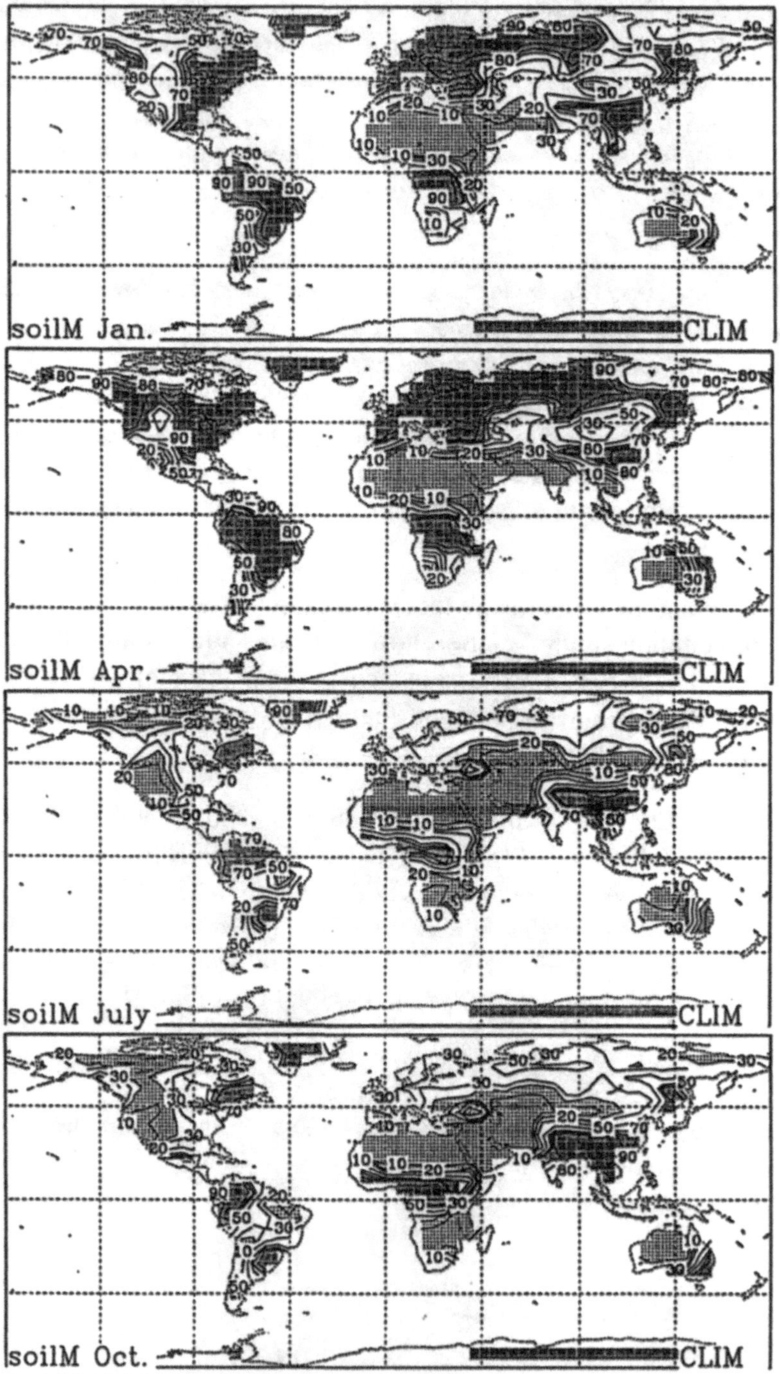

Fig.6a: Global distribution of soil wetness (100% corresponds to 0.2 m of water stored in the soil) from the climatology by Mintz and Serafini (1981)

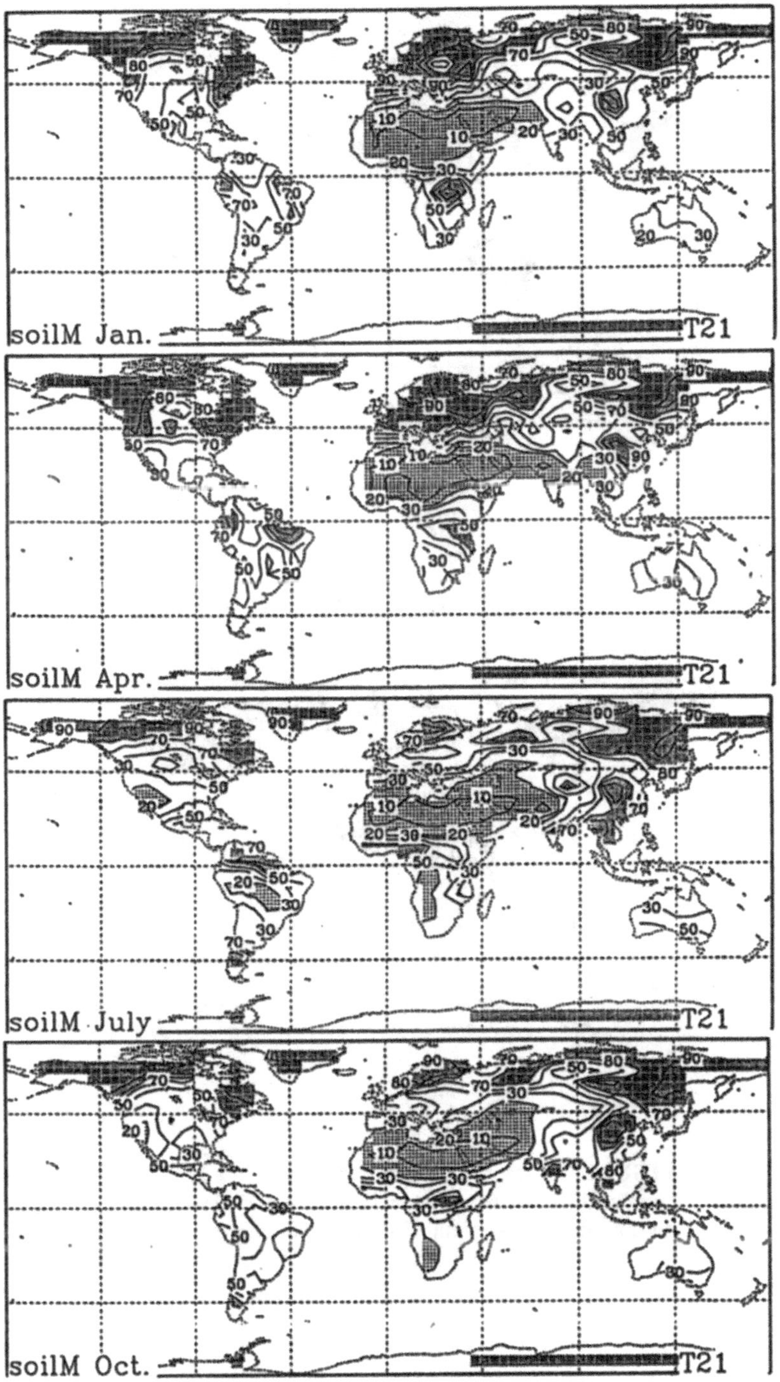

Fig. 6b as **Fig. 6a** but for the long-therm average from the 20 year integration with the ECHAM2 model (redrawn from Dümenil and Todini, 1992)

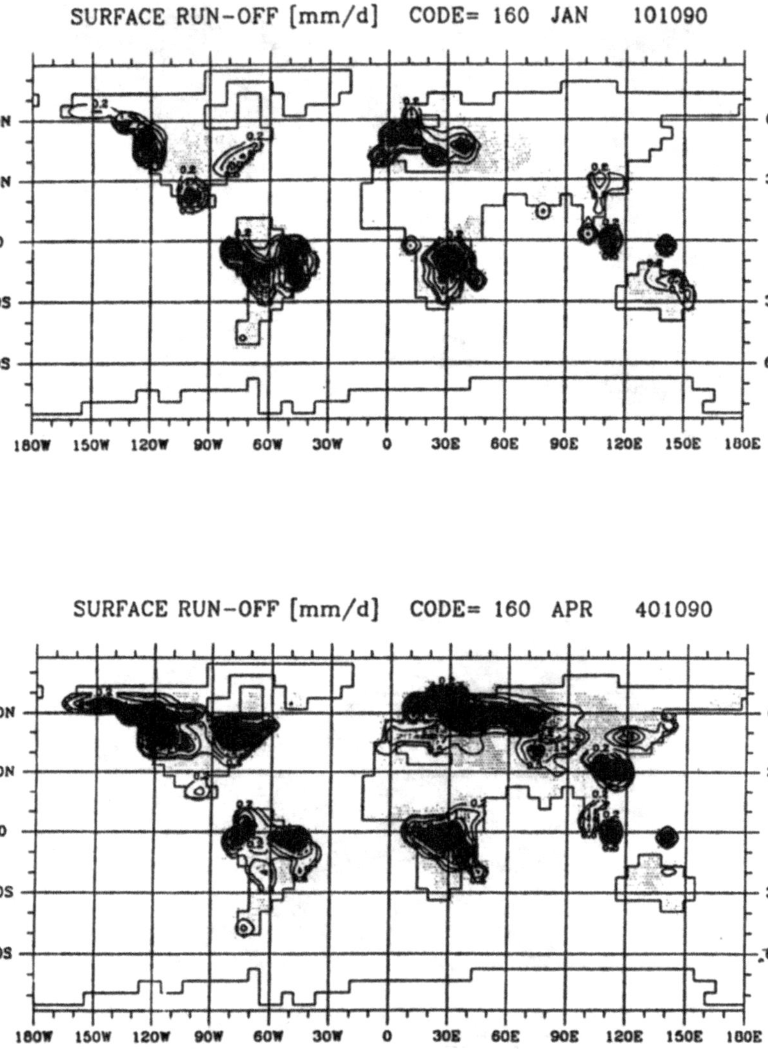

FIGURE 7a: Monthly mean local runoff in a grid cell as computed by the Arno-model (Dümenil and Todini, 1991) during the 20 year simulation with the ECHAM2 general circulation model for January and April. Contour interval is -1.

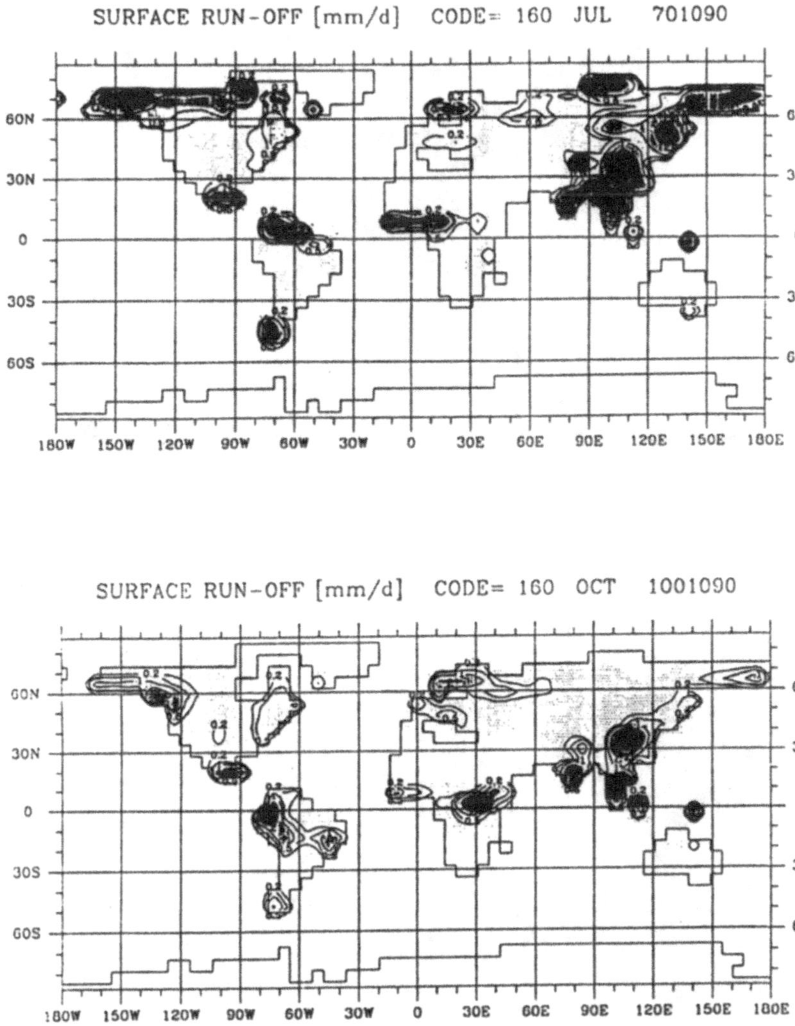

FIGURE 7b: Monthly mean local runoff in a grid cell as computed by the Arno-model (Dümenil and Todini, 1991) during the 20 year simulation with the ECHAM2 general circulation model for July and October. Contour interval is -1.

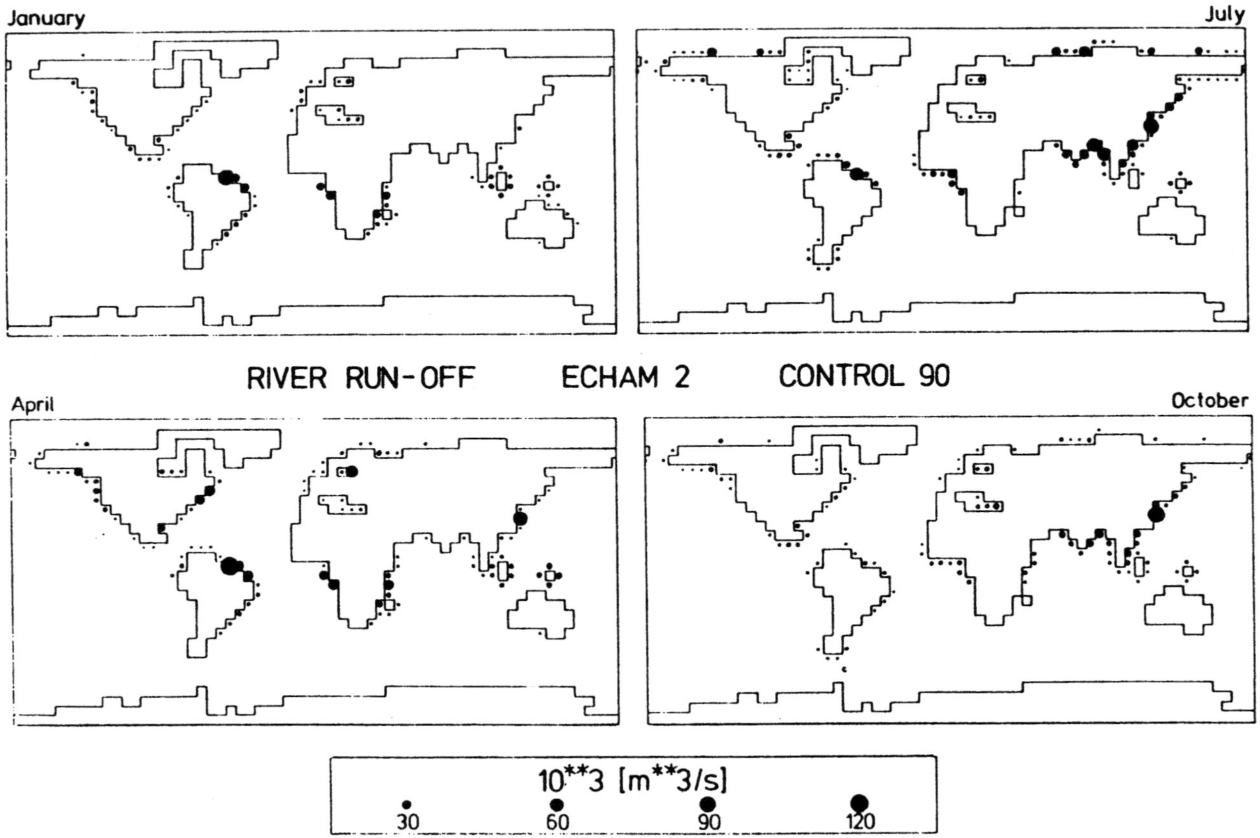

FIGURE 8: Freshwater inflow into the world's oceans during the annual cycle as simulated by the atmospheric general circulation model ECHAM2 in combination with the river-runoff model by Sausen et al., 1991.

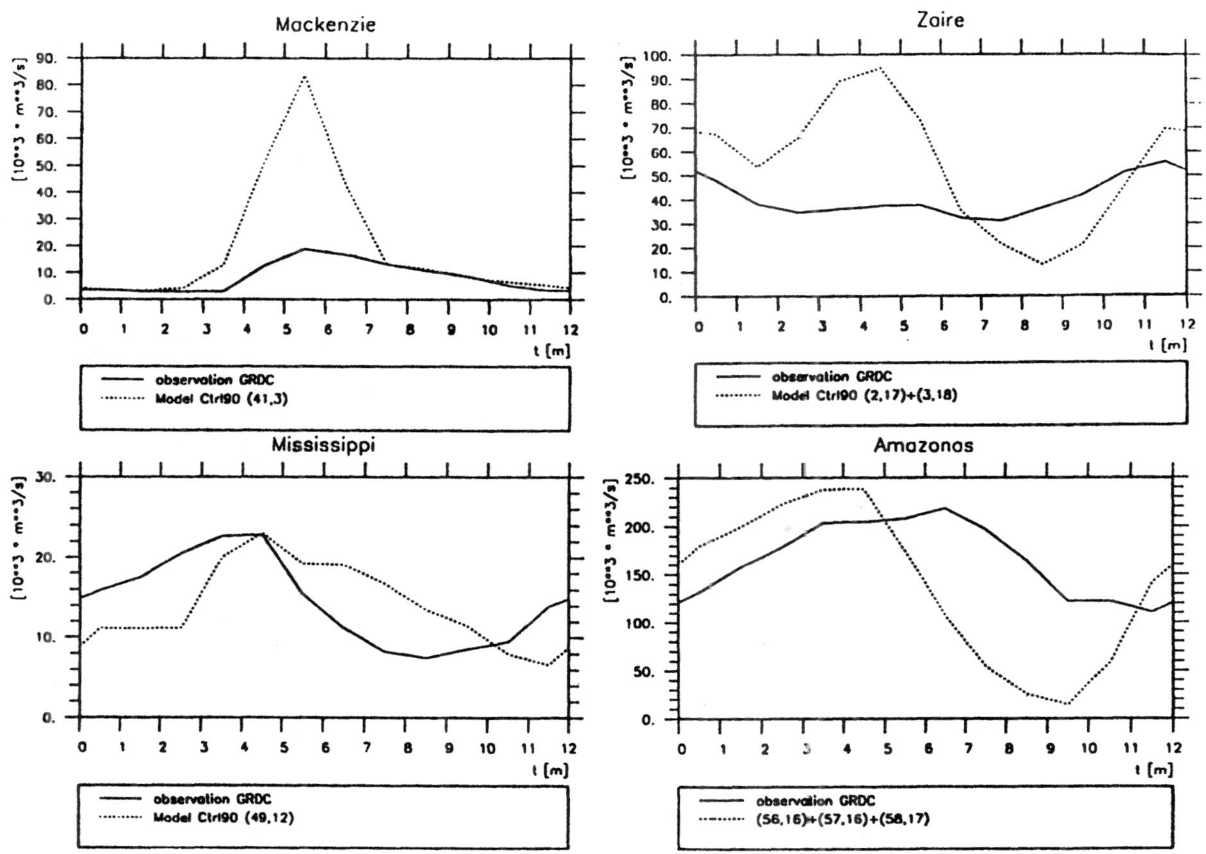

FIGURE 9: Annual cycle of freshwater inflow into the ocean collected in large river catchments for ECHAM2 model results and river gauge measurements (data courtesy of the Global Runoff Data Centre, Koblenz, Germany).

5. FUTURE DIRECTIONS

After we have outlined several aspects of problem areas concerning land surface processes in GCMs, we can now outline avenues for future research in this field.

It is obvious that the representation of land surface processes is far from adequate to answer the questions that are being posed to the climate modelling community. State-of-the-art models such as ECHAM2 require improved parameterization schemes for the processes of snow accumulation and ageing of the snow at the surface, evapotranspiration and soil hydrology. These schemes must also be able to incorporate small-scale effects and thus accomodate for heterogeneous distributions of surface quantities.

At the same time systematic errors, which are not directly due to land surface processes, will have to be minimized. One problem that needs to be resolved is, of course, the oversimplification of the description of the atmospheric flow characteristics in low resolution models. With more powerful computers some of the systematic modelling errors may be reduced by higher horizontal resolution. For the atmosphere the obvious choice is to resolve the "weather" at greater detail because climate is a longterm average of the phenomena that we experience as weather. One obvious improvement in the representation of the hydrological cycle would be to correct the distributions of precipitation, evaporation and river-runoff which are dominated by small-scale heterogeneity and variability due to precipitation systems, vegetation and soiltype distributions, etc. It appears evident that there is no realistic alternative to a significant increase in horizontal resolution in order to at least properly resolve the synoptic scale. Nesting techniques and the application of regional or mesoscale models, model output statistics and EOF analyses need to be explored to provide the small-scale information required e.g. to obtain a realistic distribution of precipitation in space and time on a sub-grid scale.

The demand for upscaling (aggregation in order to simplify the model parameterization) and downscaling (creating a distribution of heterogeneous quantities over the grid from grid area averaged information) applies to all land surface processes relevant to climate modelling. Yet they become particularly important if climate models are interactively coupled to biome models. There is a requirement that will have to be investigated in the near future to overcome the current statically defined vegetation distributions. These are presently being fixed to values which are typical of present-day climate and do not respond to any simulated climate change during an integration.

While a simplification of the physical processes is consistent with the strongly simplifying view taken in low resolution models, higher resolution models may allow a higher degree of sophistication of the representation of these processes. One could also extend the complexity and number of processes that need to be included in order to deduce the information required (e.g. impact assessment).

As important as the development of techniques to describe the physics and dynamics of the parameters associated with or interacting with land surface processes is the creation of datasets for their initialisation and validation. Field experiments from the scale of a single site to a whole grid area may be used in order to calibrate and validate one-column applications of parameterization schemes. Shuttleworth (1991) suggests that time-averaged data may be useful as a surrogate for area-averaged data to be compared with GCM output. Examples of such field experiments are HAPEX-MOBILHY, the Sahelian Energy Balance Experiment, the Amazon Region Micrometeorological Experiment, the First ISLSCP Field Experiment, the experiments at La Crau and LOTREX-HIBE (for details cf. Shuttleworth, 1991). The large-scale campaigns GEWEX and IGBP, BAHC, ECHIVAL/EFEDA and HAPEX-SAHEL are in their planning stage.

The quality and resolution requirements for datasets to be used in global models make the collection of these data an enormous task. The degree of accuracy that is required goes far beyond the standards of measurement that have been achieved so far. The large amount of data that needs to be collected at e.g. 0.5 degree horizontal resolution in order to validate the parameters of the hydrological cycle alone calls for international, i.e. global cooperation. Work has begun with initiatives such as the Global Precipitation Climatology Centre at Deutscher Wetterdienst (Offenbach, FRG) and the Global Runoff Data

Centre (Koblenz, FRG), large-scale projects such as GEWEX and IGBP being established in order to provide over the next decade the global data bases that are required for model initialisation and validation of model output.

REFERENCES

Baumgartner A, Reichel E (1975) World Water Balance: Mean annual global, continental and maritime precipitation, evaporation and runoff. Elsevier, Amsterdam.

Behr H, Dümenil L (1991) The snow climatology in decadal model simulations. Large-scale Modelling Report No. 9. Meteorologisches Institut der Universität Hamburg, Hamburg.

Blondin C (1989) Research on land surface parameterization schemes at ECMWF. Proceedings of the workshop on parameterization of fluxes over land surface, 24-26 October 1988, ECMWF, Reading, U.K.

Dickinson RE, Henderson-Sellers A, Kennedy PJ, Wilson MF (1986) Biosphere-Atmosphere Transfer Scheme (BATS) for the NCAR Community Climate Model. NCAR, Boulder, CO. Tech Note ITN-275+STR.

Dickinson RE (1989) Modelling the effects of Amazonian forestation on regional climate: A Review. Agric For Meteorol 47:339-347.

Dümenil L, Todini E (1991) A rainfall-runoff scheme for use in the Hamburg climate model. Submitted to J Hydrol.

Manabe S (1969) Climate and ocean circulation: I. The atmospheric circulation and the hydrology of the earth's surface. Mon Wea Rev 97:739-774.

Mintz Y, Serafini Y (1981) Global field of soil moisture and land-surface evapotranspiration. NASA Goddard Flight Center Techn Memo 83 907, Research Review 1980-81:178-180.

Prentice IC, Cramer W, Harrison SP, Leemans R, Monsernd RA, Solomon AM (1991) Predicting global vegetation patterns from plant physiology and dominance, soil properties and climate. In preparation.

Roeckner E, Dümenil L, Kirk E, Lunkeit F, Ponater M, Rockel B, Sausen R, Schlese U (1991) The Hamburg version of the ECMWF model (ECHAM). In: Boer GJ (ed): Research Activities in Atmospheric and Oceanic Modelling, CAS/JSC working group on Numerical Experimentation, 13, WMO-TD-No. 332, 7.1-7.4.

Sausen R, Schubert S, Dümenil S (1991) A model of the river-runoff for use in coupled atmosphere-ocean models. Submitted to J Hydrol.

Sellers PJ, Mintz Y, Sud YC, Dalcher A (1986) A Simple Biosphere Model (SiB) for use within general circulation models. J Atmos Sci 43, 6: 505-531.

Shuttleworth WJ (1991) Insight from large-scale observational studies of land/atmospherice interactions. In: Wood EF (ed) Land surface-atmosphere interactions for climate modelling Kluwer Akademic Press, Dordrecht.

REVIEW OF RECENT ADVANCES IN DYNAMICAL EXTENDED RANGE FORECASTING FOR THE EXTRATROPICS

J.F. Royer
Météo-France
Centre National de Recherches Météorologiques
42 Av G. Coriolis
F-31057 Toulouse Cédex
France

1. INTRODUCTION

Within current definitions, Dynamical Extended Range Forecasting (DERF) refers to numerical forecasts performed beyond the medium range until all dynamical influence of the initial conditions has been lost. The numerical approach to extended range predictions was formulated more than 20 years ago by Smagorinsky (1969) under the name "Deterministic Extended Range Forecasting". DP Baumhefner (personal communication) has suggested that the term "Numerical Extended Range Forecasting" would be preferable, but for this review we will use the expression "Dynamic Extended Range Forecasting" (DERF), since it has already been used in several publications. The current use of the adjective "dynamical" rather than "deterministic", or "numerical", puts more emphasis on the dynamical mechanisms by which the initial conditions can influence the time averaged properties of the flow (Shukla, 1981), and is an implicit recognition that, while the model used is deterministic, the forecasts contain a stochastic element due to the growth of the initial uncertainties. Initial conditions and boundary conditions need to be specified in order to be able to integrate a numerical model. The DERF can be considered as a transition between the two extreme kinds of predictions : deterministic short range prediction where the influence of the initial conditions is dominant, and climate prediction which is by its nature a statistical concept and where the key role is played by the boundary conditions.

If we try to define more precisely the time-range of DERF we could consider that its lower range should start when the influence of anomalous boundary conditions becomes important for the evolution of the atmospheric state, and its upper range when all influence of initial conditions has been lost. The limitations to predictability due to the amplification of uncertainties in the initial conditions have been considered by Thompson (1957). Studies of error growth in real forecasts (Lorenz; 1982) have been refined to study their scale dependence as a

NATO ASI Series, Vol. I 6
Prediction of Interannual Climate Variations
Edited by J. Shukla

function of wave-number (Dalcher and Kalnay, 1987, Boer et al, 1988). It has been established that error growth and saturation is faster in the high wavenumbers, and this progressively influences the lower wavenumbers through non-linear energy transfer. Longer influence of initial conditions can be found in the planetary scales. Michaud (1990) has estimated that the memory of the initial conditions can exceed one month. The influence of initial conditions is significant for 15-45 day means as shown by Shukla (1981), and can be still important for day 31-60 (Shukla and Fenessy, 1988). Anomalous boundary conditions have an indirect influence by heat exchange with the atmosphere (Shukla, 1984). Cubasch and Wiin Nielsen (1986) have estimated that it takes at least 10 days for the anomalous sea surface temperatures (SST) to influence the overlying atmospheric flow, and Miller and Roads (1990) that it takes about 2 weeks for instantaneous flow, and 3-4 weeks for time-averaged predictions. With these bounds one can consider that the time range of the DERF is comprised between 10-30 and 10-60 days. The essential difference between medium range forecast (MRF) and DERF is that the stochastic character of the forecast becomes more prominent as the time range is being extended.
Dynamical forecasting beyond the limit of deterministic predictability is viable only if some statistical postprocessing of the times series generated by the models is performed. The DERF is thus one of the most complex and challenging forecasting problems as it requires global analyses for the specification of observed initial and boundary conditions, the use of sophisticated numerical models, a statistical treatment of the model results, and a verification of the forecast results. Most DERF studies have been generally restricted to the extratropics, although the global models used to perform the extended forecasts naturally include the tropical regions. Forecasting for the tropics can be considered as a specific problem since the space and time scales, and the instability mechanisms, of tropical phenomena are completely different from those in midlatitudes.

The basic concepts and developments of DERF have been documented in several review papers (Barnett and Somerville, 1983; Kalnay and Livezey,1985; Miyakoda, 1985; Shukla, 1985; Hollingsworth et al, 1987). The present paper will attempt to summarize the results of DERF studies which have been published in the last 5 years. One of the most significant factor for the development of DERF in recent years has been the gradual increase in computational power that has made possible more extensive application and

Table 1 : Synopsis of various DERF experiments published in the Last 5 years

Center	References	Horizontal/Vertical resolution	Time range (days)	Number of cases	Period	Ensemble size	Climatology
ECMWF	Cubasch, Wiin-Nielsen 1986	T63/42/21 L16	60	2+3	JF79+JF83	1	
	Molteni et al 1987	T42/21 L16	30,60	4+38	Dec82-Feb85	9 LAF	ECMWF (79-83)
	Palmer et al. 1990	T106 L16/19	30	42	Apr 85-Sept 88	2 LAF	
	Tibaldi et al. 1990	T106/63/42/21 L19	"	24	DJF 85-90	2 LAF	
	Brankovic et al. 1990	T63 L16/19	"	16	Apr 85-March 86	9 LAF	ECMWF (79-84)
UKMO	Mansfield 1986	Hem. 330km L5	50	71	Dec 74-81.	1	
	Owen, Palmer 1987	2.5 x 3.75° L11	90	3x2	DJFM 82/83	1	
	Murphy,Dickinson 1989	" "	40	64	1982-87	1	1951-80
	Murphy 1990	" "	30	8	Dec 85-Sep 87	7 LAF	NORMALS
	Milton et al 1991	2.5 x 3.75° L15	30	2Xmonth	since Dec. 88	9 LAF	
GFDL	Miyakoda et al. 1986	N48 (1.875°) L9	30	8	Jan. 77-83	3 different analyses	OORT
	Sirutis, Miyakoda 1990	" x 4 different	"	"	"		
	Miyakoda, Sirutis 1990	" physics	"	"	"	"	
NMC	Tracton et al. 1989	R40(+some T80) L18	30	108	14/12/86-01/04/87	contiguous	NMC(78-85)
	Roads 1989	" "	"	"			
	Saha et al 1990	T40 "	90	128	03/05/90-06/12/90	"	
JMA	Yamaha et al 1991	T63 L21	35	8	JF 87-90	9 LAF	NMC+JMA (51-80)
COLA	Shukla, Fennessy 1988	4 x 5° L9	60	4	DJ 82/83	3 LAF	NMC(JF 79-84)
	Fennessy, Shukla 1991		"	4 + 2	" + JF 87		(excl 1983)
CNRM	Déqué 1988	T21 L15	45	5	Jan. 80-84	9 LAFxMC	ECMWF (79-89)
	Déqué 1990	T42 L15	"	1	Jan. 88	5 LAF	(excl. year
	Déqué 1991	T42 L20	"	21	DJF 83-90	5 LAF	of forecast)
LMD	Michaud 1990	64x50 grid L11	45	5 x 4	DJ 82-83	5 LAF	ECMWF (DJ79-85)
					+ DJ 79-85	+1	(excl.DJ82/83
NCAR	Baumhefner 1991	T31 L12	30	49	DJF 79-90	10 MC	NMC (79-85)
CCC	Boer et al. 1988	T20 L10	30	42	DJF 79-86	6 LAF	NMC (79-86)

<u>Notes</u> : **Horizontal resolution** : grid size given by increments of latitude and longitude in ° for grid point models with regular latitude-longitude grid, or by the letter T (triangular) or R (rhomboidal) followed by the size of the truncation for spectral models. (Hem. : Hemispheric: for the LMD model the number of grid points is given since the grid has constant increments in sine of latitude).- **Ensemble size** : LAF (Lagged Average Forecasts); MC (Monte Carlo)

testing of numerical models in extended range prediction experiments with real data. Also, systematic errors of numerical models used in DERF studies have been largely reduced due to improvements in their physical and numerical formulations. In turn, these improvements have prompted some forecasting centers to use the DERF as a tool in studying model climate drift. The DERF experiments have been performed either by extending beyond the medium range the operational forecasts performed operationally in several centers (ECMWF, NMC, JMA), or by initializing climate models with the global analyses obtained from the operational centers (Table 1). Experiments using a "perfect model" approach (e.g. Tribbia and Baumhefner, 1988; Murphy JM, 1988) have not been included in this table.

2. FORECAST VERIFICATION

As pointed out by Murphy and Winkler (1987) forecast verification should ultimately be based on the joint distribution of forecasts and observations, and any single score to measure the skill is necessarily incomplete. However, standard verification scores are usually restricted to a measure of distance between forecast field and observed (mean square error MSE, or it square root RMSE), or a measure of similarity of patterns like the anomaly correlation coefficient (ACC). AH Murphy (1988) and Murphy and Epstein (1989) have derived the analytical relationship between RMSE and ACC, showing that the correlation ignores different biases, and that its square is thus a measure of potential skill, rather than actual skill. A criterion for defining the limit of practical predictability has been proposed by Saha and Van den Dool (1988). The field that is traditionally used in most publications for verification of DERF forecasts is the 500 hPa geopotential height. Other fields have been also occasionally used such as 300 hPa (Shukla and Fennessy, 1988; Fennessy and Shukla, 1990) or 1000 hPa geopotential height (Miyakoda et al, 1986), sea level pressure (Murphy, 1990), 1000-700 hPa thickness (Roads, 1989), 850 hPa temperature (Palmer et al, 1990a; Tibaldi et al, 1990; Déqué, 1990; Royer and Déqué, 1990), or geopotential and temperature at 850- 500- and 200 hPa (Déqué, 1988).

Skill scores and ACC are dependent on the climatology used. This makes the comparison of the skill of different models, which use observed climatologies based on different analysis systems and different periods, difficult. Murphy and Dickinson (1989) and Murphy (1990) showed that the use of normals based on too short data sample may usually overestimate the skill, in particular if they are based on 10 years or less, due to climate sampling error. However the differences between climatologies based on different analysis systems and different years can also introduce large fluctuations, and the use of a climatology inconsistent with the forecasts may produce spuriously high ACCs (Molteni et al, 1987; Boer et al, 1988). Large differences in ensemble mean correlations were obtained by Barker and Horel (1989) simply by varying climatology and treatment of systematic errors. They attributed the lower skill reported by Roads (1989), compared to that of Tracton et al (1989), to the use of a climatology that removes low-frequency variations. Several methods have been used for averaging the ACC scores in a series of experiments. The simplest and most frequently used method is just to take the arithmetic mean of the ACC obtained in each experiment. Another method is to average the Fisher Z-transform of the correlations, which resemble more closely a normally distributed population. For ease of interpretation the ensemble averages of the transformed correlations are usually transformed back to obtain the ensemble mean anomaly correlation (Branstator, 1986; Roads, 1989; Palmer and Tibaldi, 1988; Miller and Roads, 1990; Barker and Horel, 1989). The transformed correlations can also be used more easily to conduct significance tests (Chen, 1990) and to estimate confidence intervals. A third method, used for example by Van den Dool and Saha (1990), is to compute a mean correlation after averaging the moments (variance and covariance). This has the advantage of giving more weight to cases with large amplitude and has a simple geometrical interpretation in multi-dimensional phase space as equal to the cosine of the angle between the 2 vectors representing predicted and observed anomalies.

3. POSTPROCESSING THE RESULTS

Since the aim of the DERF is to predict changes in weather regimes rather than instantaneous weather, various techniques have been proposed for filtering the unpredictable components due to individual synoptic disturbances: time averaging, spatial averaging and ensemble averaging. Time averaging is perhaps the most obvious and widely applied method to eliminate high-frequency baroclinic waves and isolate the potentially predictable signal found in the slowly varying components of atmospheric circulation, such as stationary and low-frequency planetary waves. However the choice of the most appropriate averaging period is not obvious and different averaging periods have been used in various DERF experiments. Miyakoda and Sirutis (1985) have found that the use of a 10-day low-pass filter is adequate for the removal of unpredictable high-frequency components associated with baroclinically unstable waves, which correspond to a substantial peak in the frequency range of 2-7 days. The application of 10-day means increases the predictability of the AC=0.5 criterion for about 1.5 days (Tracton et al, 1989). Monthly means (day 1-30) are frequently used since it corresponds to traditional practice and user needs. Other averaging periods such as 5-day means, 15-day means or even 20-day means are also occasionally used. In general, the ACC of time averaged means is larger than the average of daily scores and improves with the length of the averaging period. The pros and cons of time averaging have been discussed by van den Dool (1985): time averaging can only help to bring out more clearly the skill of the daily forecasts by improving the signal to noise ratio. One of the consequences of temporal smoothing is the suppression of small scale variability. Such effect can be also obtained directly by truncating these scales in spectral space, such as retaining only planetary waves of wave number 0-4 (Shukla, 1981) or 1-3 (Mansfield, 1986). Dalcher et al (1988) "tempered" ordinary dynamical forecasts, damping their amplitude by wavenumber, with good results. Projection into empirical orthogonal functions (EOF) has also been used to remove higher frequency components and to identify weather regimes. The first few EOFs represent the dominant low frequency patterns and the most persistent anomalies which may be more predictable. Tracton et al (1989) used a projection on the first 16 EOFs, and Palmer et al (1990) on 3 rotated EOFs. A projection in the phase space spanned by 2 principal EOFs was used by Brankovic et al (1990) to study the trajectories and dispersion of ensemble forecasts. They also proposed the usage of clustering technique as a probalistic way to extract the most useful information from an ensemble of forecasts.

4. ENSEMBLE TECHNIQUES

Leith (1974) proposed the use of ensemble forecasting as a practical approximation to the stochastic-dynamic prediction framework (Epstein, 1969) developed for estimating the probability distribution of atmospheric variables. Several methods can be used for generating an ensemble of initial states : perturbation by random errors of analysis known as Monte Carlo (MC) forecasts (used for example by Baumhefner, 1991); using different analyses (DA) based on different models or systems (Miyakoda et al,1986; Miyakoda and Sirutis, 1990; Kalnay and Dalcher, 1987) which reflect data uncertainty; using of successive analyses from the same system, proposed by Hoffman and Kalnay

(1983) under the name Lagged Average Forecasting (LAF). The LAF method has been used in most of the recent ensemble DERF experiments.
In the case of Monte Carlo forecasts the mean sampling error variance is inversely proportional to the size of the ensemble. But for LAF an increase of the number of forecasts has the detrimental effect of including forecasts from earlier initial conditions, so that there may be an optimal size. Kalnay (1990) suggested a simple improvement upon LAF that eliminates this problem by allowing the generation of additional LAF members with approximately the same short "equivalent age" with respect to the latest forecasts. LAF has, on average, less skill than the last forecast of the ensemble up to about day 10, but in the extended range, when individual ensemble members become more equally likely, LAF has a positive effect. Miyakoda et al (1986) found that the ACC for the ensemble average are slightly better compared with the arithmetic average of individual coefficients. LAF has a stronger impact on MSE criterion than ACC (Brankovic et al.,1988). Brankovic et al (1990) found the main benefit to be for days 11-20, but not for ACC for days 21-30. Tracton et al.(1989) and Yamada et al (1991) found that for the first ten day mean (1-10) the impact of LAF is small; for the third 10-day mean they find a significant impact of LAF with a improved skill when the ensemble size is increased up to 9, most of the improvement being already obtained with a 3-member ensemble. In a series of 45-day predictions with the French model, Déqué (1991) showed that the improvement of an ensemble of 3 forecasts versus a single forecast is about half the expected improvement analytically extrapolated for an infinite ensemble. The impact of ensemble averaging is larger when the mean skill of individual forecasts is better, whereas averaging of forecasts of poor quality does not increase the skill as observed by Tracton et al (1989). Roads (1988) has found that optimal weighing of LAF can lead to increased ACC scores in idealized experiments, and Dalcher et al (1988) also found improved LAF forecasts when appropriate weights were used. However, Murphy (1990) could not confirm such impact in real forecasts, and has shown that this can be attributed to the external component of error growth, since a much larger increase in skill is achieved by ensemble averaging for perfect model scores than for real predictions. His interpretation is that ensemble averaging improves skill through the reduction of random forecast errors arising from initial state uncertainty, but does not affect the errors caused by model deficiencies. Brankovic et al (1990) also concluded that the presence of model errors can reduce considerably the gain from ensemble averaging.

5. CORRECTION OF MODEL SYSTEMATIC ERRORS

The amplitude of the forecast error in a numerical forecast may depend on several factors : the distribution of the error in the initial field, the dynamical instability of the flow and the model-generated errors due to deficiencies in its formulation (Molteni et al, 1987). The influence of the increase in horizontal resolution on model systematic errors and skill scores have been analysed by Tibaldi et al. (1990) in a set of extended range integrations with ECMWF model. The similar behaviour of systematic errors for resolutions higher than T42 led them to the conclusion that T42 can be a satisfactory compromise for many DERF studies, whereas the different behaviour of T21 and its poorer skill scores made its use questionable. The impact of different physical parameterizations has been investigated by Sirutis and Miyakoda (1990) who have shown that the improvement of physics had more impact on performance in the medium-range, but that 30-day means appeared more sensitive to circulation regime in the initial conditions.

In spite of continued efforts for their reduction through model improvements, the errors due to climate drift of the model are still large enough to degrade the performance of monthly forecasts. Boer (1991), on the basis of the budget equations he has developed, has shown that the random error does not interact very strongly with the systematic error, so that it should be possible to "subtract out" of the systematic errors. Removing systematic errors a posteriori by means of an empirical adjustment can improve the skill scores of time averages. Much larger improvements in skill occur when the corrections are computed on the same period as the forecast (dependent data) (Miyakoda et al, 1986; Baumhefner, 1988; Murphy, 1990). For operational use the systematic errors have to be estimated a priori from independent data by computing the climatology of the model on an independent limited sample of forecasts used as a training period. Such independent correction leads to much lower skill improvements than the dependent correction (Murphy and Dickinson, 1989; Yamada et al, 1991). In the case of the NMC DERF experiment Tracton et al (1989) found a small improvement of the 1-30 day ACC from 0.39 to 0.45. Barker and Horel (1989) showed that whereas the skill of instantaneous forecasts is not much affected by the removal or retention of systematic errors, the skill of time averages that include low-frequency variability may be enhanced by systematic error correction. Déqué (1991) has developed generalized analytical expressions based on those of JM Murphy (1988, 1990) to extrapolate the decrease of RMSE, and the increase of the ACC, as a function of the number of independent reference integrations used to compute the model bias. He has shown that the correction of the empirical systematic error improves the model skill only if a large enough number of reference integrations is used for its estimation, and observed that the asymptotic values were practically reached with 10 reference simulations. The subtraction of an empirical bias removes a part of the systematic error, but adds a random error which is function of the number of reference simulations and of the interannual variability of the model error. This empirical correction method does not take into account the possibility pointed out by O'Lenic and Livezey (1988) that the systematic error may depend on the initial observed low-frequency circulation. Such a stratification of the systematic error would require a much large sample of DERF forecasts with the same model than is currently available (Baumhefner, 1988).

Another possibility to reduce the model systematic biases is to introduce correcting forcing terms into the model equations during the course of the integration (within forecast correction) to counteract the model's initial drift. Johansson and Saha (1989) have shown with a model based on the equivalent barotropic equation with forcing and dissipation that the recurrent corrections are able to improve the transient energy simulation and decrease the growth rate of random error. In climate simulations with a low resolution (T21) version of the Hamburg GCM Sausen and Ponater (1990) have confirmed that a reduction of model drift by constant correcting terms can improve not only the equilibrium climate of the model, but also its transient behaviour, allowing to simulate more realistically the variability and the frequency of blocking events. The main drawback of this statistical correction method, which has up to now prevented its use in DERF experiments, is that it would require a frozen version of the model, and a training sample of cases, large enough to allow a stable estimation of the correcting terms.

6. INFLUENCE OF BOUNDARY CONDITIONS

Surface boundary forcing can be one of the major sources of climatic predictability at the monthly to seasonal time scales. The possible role of the inclusion of the anomalous forcing in boundary conditions in enhancing the predictability of monthly and seasonal averages considered by Shukla (1984) has been investigated in several experiments. There is now a growing consensus that the forecasts can be improved using the observed rather than the climatological Sea Surface Temperature (SST) data. The usual method used in DERF experiments is to use a persistence of the observed initial SST anomalies.

Cubasch and Wiin-Nielsen (1986) have concluded from 60-day prediction experiments at various resolutions with the ECMWF model with climatological and observed SST, that a more realistic SST distribution may have a beneficial impact on the quality of the forecast for extended range predictions. The influence of anomalous boundary forcings depends not only of their magnitude and spatial scale, but also of the structure of the dynamical flow interacting with the lower boundary. Mansfield (1986) comparing the effect of extratropical versus tropical SST anomalies, found only a small and inconsistent effect for extratropical anomalies, whereas, on the contrary, tropical SST anomalies had a significant impact on midlatitude circulation. The experiments of Owen and Palmer (1987) and Baumhefner et al (1988) have confirmed that the average forecast skill for the extratropics is generally, but not always consistently improved by using the global observed SST. The 1982/83 El Nino, one of the largest ever observed, has been used in several DERF studies (Owen and Palmer, 1987; Shukla and Fennessy, 1988; Michaud, 1990). Shukla and Fennessy (1988) showed that the use of the observed SST anomalies over the Pacific produced a very large impact on tropical prediction of circulation and rainfall for 30 and 60 day averages, and also a modest improvement over the midlatitudes, which was further enhanced by removing the climate drift of the model. Fennessy and Shukla (1991) compared the 1986/87 El-Nino more typical in magnitude, and showed that the GCM was able to simulate the difference between the 2 El Nino years. This impact was stronger on 30-60 day means than 1-30 day means. Mo et al (1991) showed that SST anomalies were responsible for about half of the high skill in the monthly forecast for June 1988. Mo and Kalnay (1991) performed experiments showing that the use of SST anomalies could have a large impact on mid-latitude forecasts if the SST anomalies were the cause and not the result of atmospheric anomalous forcing. Miller and Roads (1990) found that the coupled model is better than the model with climatological SST for predictions of 30-60 day averages, the improvement on skill increasing steadily from 1-day to 90-day averages. For a non El Nino year Déqué(1990) found that the use of actual instead of persistent monthly mean SST produced a significant increase of skill for the divergent tropical circulation; the use of daily observed instead on monthly mean SST does not improve this skill, and that there is no significant impact on geopotential field in midlatitudes. The impact of continental snow cover in 2 experimental regions on 30-day forecast was investigated by Walsh and Ross (1988) who showed that the large-scale circulation is more sensitive to Eurasian snow cover than North American snow cover.

All these studies confirm that a correct inclusion of boundary conditions of global SST, snow and sea ice, can enhance the skill of predictions of the space-time averaged anomalies at the monthly to seasonal scale.

7. CURRENT LEVEL OF SKILL

Typical values of ACC coefficients for 10- and 30- day means are shown in table 2. One can see that in spite of differences in the experiments, the level of skill of the different models is comparable, and that the average skill is marginal beyond 10 days, and almost negligible beyond 20 days.

In fact much of the skill in 1-30 day mean reflects the influence of the first part of the extended runs. Roads (1988) has suggested that it may be preferable to use a time averaging at less than 30 days to eliminate the influence of the non-skillful later part of the integration. Tracton et al. (1989) have found that the best estimate of the complete 1-30 day mean in the NMC DERF experiment was provided by averaging the first 7-8 days of the integration (increase of correlation from 0.39 to 0.47). Miyakoda and Sirutis (1985) excluded the first 10-day that have obvious skill and based most of their analysis on days 10-30 for their 8 January cases with the GFDL model. The skill of time averaged forecasts remains higher than the skill of persistence until about 30 days. There is a large sensitivity of the forecasts to initial conditions (Sirutis and Miyakoda, 1990). Baumhefner (1991) has found an average range of 0.30 in ACC among the 10 different members of a Monte Carlo ensemble. A large case to case and interannual variability of the skill has been noted in all experiments. The large temporal fluctuations of the skill are particularly apparent in the series of contiguous forecasts of the NMC DERF experiment (Tracton et al, 1989), where the low frequency variations of the skill suggest its possible connection with weather regimes (Tracton, 1990).

There are also seasonal variations of skill and it has been observed that skill is generally higher in winter and spring, and lower in summer (Murphy and Dickinson, 1989; Murphy, 1990; Brankovic et al, 1990; Yamada et al, 1991). This can be explained by the fact that predictability time is longer for large-scale quasi-stationary waves that have their largest amplitude during winter, whereas during summer less predictable smaller scale flows tend to prevail (O'Lenic and Livezey, 1989)

Table 2 : Example of 10-day and 30-day mean Northern Hemisphere 500 hPa height anomaly correlations in some recent DERF experiments

		Forecast period			
		1-10	11-20	21-30	1-30
Tracton et al. (1989) winter 1986/87		0.72	0.18	0.12	0.39
Murphy (1990)	(*)	0.7_{0}	0.1_{6}	-0.0_{4}	0.4_{2}
Palmer et al. (1990)	(*)				
(oct-march) 85/86		0.8_{2}	0.2_{0}	0.1_{3}	0.4_{0}
86/87		0.7_{8}	0.0_{6}	0.0_{4}	0.3_{2}
87/88		0.7_{8}	0.2_{1}	0.1_{0}	0.4_{2}
Yamada et al. (1991)					
Winter		0.83	0.32	0.22	0.53
Spring		0.74	0.07	-0.04	0.17
Summer		0.70	0.10	-0.01	0.29
Autumn		0.73	0.11	0.02	0.28

(*) : The smaller last figure is not reliable since these values were obtained visually from fig.4 of Murphy and fig.11 of Palmer et al..

Variability in skill is related to the rate at which initial errors grow, and atmospheric instabilities which are the basic cause of forecast error growth have strong geographical variations. One should therefore expect to find some regional dependence of skill. Maps showing the regional variability of forecast skill have been shown in several experiments (Miyakoda et al, 1986; Boer et al., 1988; Murphy, 1990; Royer and Déqué, 1990) but this seems to show only loose correspondence with the results of potential predictability studies (Shukla and Gutzler, 1983; Chervin, 1986; Zwiers, 1987). Chen (1990) has found that during ENSO winters the average forecast skill over the Pacific sector is significantly higher than over the Atlantic sector, where a much larger variability of skill is found. Palmer (1988) found a strong relationship between flow regimes over the Pacific/North American region and model skill.

8. FORECASTING THE SKILL

Due to the general low level of skill and considerable case-to-case variations in predictive skill in the extended range, it has been stressed that an essential requirement for the application of DERF is to be able to diagnose, understand

and predict the forecast skill of individual forecasts, in order to discriminate a priori the forecasts that can be of practical use. It has been noted that more skillful forecasts tend to be associated with persistent situations in which the atmosphere is more stable (Mansfield, 1986; Palmer, 1988; Palmer and Tibaldi, 1988), and Chen (1989) has suggested the usage of forecast persistence as a predictor of skill for the medium range. Forecast quality may depend on the flow regime (Palmer and Tibaldi, 1988). The intensity of the jet streams seems a crucial factor for determining the large-scale flow pattern. Above normal zonal flows have been generally found to be associated with higher predictability (Horel and Roads, 1988; Yamada et al, 1990), whereas the onset of major blocking events coincide with large decrease in skill (Tracton et al., 1989). This may be due in part to the inability of the current models to reproduce the appearance of a blocking event not present initially, and to simulate properly its evolution. The simulation of blocking activity is actually one of the main problems for extended and long range prediction and a large portion of the systematic error is associated with the failure of blocking forecast (Miyakoda and Sirutis, 1990). Prediction of blocking is generally not satisfactory even in the medium range (Tibaldi and Molteni, 1990), but can be sensitive to changes in model formulation, and can be improved in some cases with increased resolution (Tracton, 1990) or with more elaborate physical parameterizations (Miyakoda and Sirutis, 1990; Palmer at al, 1990a).

Palmer (1988) showed that the skill of numerical forecasts is strongly influenced by the sign and amplitude of the Pacific North American (PNA) mode. His explanation, tested with a barotropic model, is that the initial error growth, which is dependent on the barotropic stability of the forecast flow, is stronger for basic states with a negative PNA index (positive height anomaly over the Aleutian islands). A more complete analysis using EOF projections of initial conditions or forecast was given by Palmer and Tibaldi (1988) for the medium range. Palmer et al (1990a) report a correlation of 44% between ACC skill score at day 11-20 and a rotated EOF1 identified as a positive PNA pattern. A general conclusion emerging from studies at medium and extended range (O'Lenic and Livezey, 1989; Tracton et al., 1989; Molteni and Tibaldi, 1990) is that the most skillful forecasts are generally associated with persistence of, or transitions to positive PNA index, while poorer forecasts occur when the PNA becomes negative. However a statistical scheme to predict skill in the medium range based on a number of predictors including indices of large scale regime transition gave only correlations of the order of 40% between predicted and actual skill scores at the end of the medium range period (Molteni and Palmer,

1991). This shows that only a small part of the variability in forecast skill can be anticipated from the values of the PNA index in the initial circulation regime.

One of the motivations for the use of ensemble techniques is that the dispersion of individual forecasts from the ensemble mean could be used as an indication of the reliability of the forecasts. The spread of multiple forecasts has indeed shown promising results in the short range (Kalnay and Dalcher, 1987) and medium range (Palmer and Tibaldi, 1988; McCalla and Kalnay, 1988). Tracton et al (1989) also found correlation between forecast agreement and forecast skill on the monthly averages, while Murphy (1990) has found that the correspondence between agreement and skill is strong only in the early part of the forecasts, and vanishes beyond 10 days. Brankovic et al (1990) have shown that the positive correlation they found between RMS spread and skill in the extended range came mainly from the impact of the annual cycle on both quantities, and that in the winter period no clear spread/skill correlation could be found. Up to now the different measures of spread have failed to provide consistent information for the prediction of skill beyond the medium range. Such failure has been partly attributed to the existence of systematic model errors and to the sampling fluctuations due to the insufficient size of the ensembles for a reliable estimation of second moments. However, even with a large number of Monte Carlo ensembles of perfect model forecasts, Barker (1991) has found that the correlation between spread and forecast error tends toward zero at long forecast times. Attempts to improve correlations between spread and skill by optimal perturbations to the analyses in regions of maximum dynamical instabilities are in progress at ECMWF (Molteni et al, 1991).

9. CONCLUSIONS AND PERSPECTIVES

All DERF experiments have shown that there is on average only marginal skill beyond 10 days. However the anomaly correlation coefficients for time averages remain in most cases positive and better than persistence until about 30 days. This means that the DERF forecasts do produce some kind of improved information relative to climatology and persistence that could explain a small fraction of the variance if used optimally in a statistical scheme in conjunction with other predictors. Even such a marginal improvement could be of practical value since the skill of other long-range forecasting techniques is also currently very low and highly variable (Livezey, 1990) However, considering the low skill and the large spread of the results of ensemble forecasts beyond the medium range, it would seem more natural and appropriate to seek to formulate and verify extended range forecasts in a probabilistic framework, as attempted by Murphy JM (1988, 1990) and Déqué (1988). Combination of ensemble forecasts from different models could perhaps provide a more realistic estimate of forecast uncertainties (Royer and Déqué, 1991). The current size of ensembles (less than 10 members) which seems adequate for an estimation of the ensemble mean, will have to be increased by an order of magnitude for a more reliable estimation of the higher moments of the probability distribution of the ensemble (such as the spread) which are required to formulate probabilistic forecasts and estimate confidence limits. A cluster analysis technique has been proposed as a way to give probabilities of possible developments (Brankovic et al, 1990). However a probabilistic forecasts of regimes is not yet likely to be successful because of the systematic biases of the current models that make them unable to simulate correctly the different circulation regimes (Miyakoda and Sirutis, 1990). A reduction or correction of the climate drift of the model is therefore essential for the improvement of extended and long-range forecasts. There is in particular a need to improve the simulation of blocking processes which are responsible for a large proportion of systematic errors, and also the simulation of tropical low frequency variability (Ferranti et al, 1990). Improvements in the specification, and ultimately the forecasting by coupled models, of external forcings such as SST, soil moisture and snow cover are also required since they become increasingly important as the range of the forecasts is extended. A larger set of experiments over different years will be necessary in order to establish the statistical significance of the skill improvements that can be expected, and for a stratification of skill variability with geographical location and weather regime that will be necessary for DERF forecasts to become practically beneficial. Progress in DERF can be a critical test to establish and strengthen the credibility of the numerical models that are increasingly applied for global change studies.

ACKNOWLEDGEMENTS: I wish to thank J. Shukla for his invitation to this Workshop. I gratefully acknowledge the valuable comments and suggestions given by David Baumhefner, Cedo Brankovic, Michel Déqué, Huug van den Dool and Eugenia Kalnay, which have greatly helped to improve the manuscript. Research on DERF at CNRM is supported in part by PNEDC (Programme National d'Etude de la Dynamique du Climat) and by the Climate Programme of the Commission of the European Communities (Contract EV4C-0066-F)

REFERENCES

Barker TW (1991) The relationship between spread and forecast error in extended-range forecasts. J of Climate 4: 733-742

Barker TW, Horel JD (1989) The impact of climatology and systematic errors upon the skill of DERF forecasts. Mon Wea Rev 117: 2835-2842

Barnett TP, Somerville RCJ (1983) Advances in short term climate prediction. Rev Geophys Space Physics 21: 1096-1102

Baumhefner DP (1988) The structure of systematic error and its influence on extended range prediction using the NCAR T31 climate model. Proceedings of the Workshop on systematic errors in models of the atmosphere (Toronto, Canada, 19-23 Sept. 1988). CAS/JSC Working Group on Numerical Experimentation, Report n° 12, WMO/TD n°273: 86-93

Baumhefner DP (1991) Evaluation of long-range forecast skill from Monte-Carlo ensemble integrations of a low-resolution NCAR Community Climate Model. Extended abstracts submitted to the ICTP/WMO international technical conference on long-range weather forecasting research (Trieste, Italy, 8-12 april 1991), WMO/TD n° 395: 15-19

Baumhefner DP, Tribbia JJ, Blackmon ML (1988) The influence of specified sea surface temperature and initial condition uncertainty on Monte Carlo extended range forecast ensembles. In : Modelling the sensitivity and variations of the Ocean-Atmosphere system (Report of a Workshop at the ECMWF, 11-13 may 1988). WCRP- 15, WMO/TD n°254: 3-10

Boer GJ, Zwiers FW, Chan E (1988) Dynamical extended range forecasting at the Canadian Climate Center. Proceedings of the ECMWF workshop on predictability in the medium and extended range, 16-18 May 1988. ECMWF, Shinfield Park (UK), p 135-152

Boer GJ (1991) Systematic and random error in an extended-range forecasting experiment. (submitted to Mon Wea Rev)

Brankovic C, Palmer TN, Molteni F, Tibaldi S, Cubasch U (1990) Extended-range predictions with ECMWF models: Time-lagged ensemble forecasting. Q J R Meteorol Soc 116: 867-912

Branstator G (1986) The variability in skill of 72-hour global-scale NMC forecasts. Mon Wea Rev 114: 2628-2639

Chen WY (1989) Another approach to forecasting forecast skill. Mon Wea Rev 117: 427-435

Chen WY (1990) Interannual variability of skill of NMC medium-range forecasts over the Pacific/North America sector. Mon Wea Rev 118: 179-188

Chervin RM (1986) Interannual variability and seasonal climate predictability. J Atmos Sci 43: 233-251

Cubasch U, Wiin-Nielsen AC (1986) Predictability studies with the ECMWF spectral model for the extended range: the impact of horizontal resolution and sea surface temperature. Tellus 38A: 25-41

Dalcher A, Kalnay E (1987) Error growth and predictability in operational ECMWF forecasts. Tellus 39A: 474-491

Dalcher A, Kalnay E, Hoffman RN (1988) Medium range lagged average forecasts. Mon Wea Rev 116: 402-416

Déqué M (1988) The probabilistic formulation: A way to deal with ensemble forecasts. Ann Geophys 6: 217-224

Déqué M (1990) Impact of prescribed sea surface temperatures on extended range forecasting. J Marine Systems 1: 61-70

Déqué M (1991) Removing the model systematic error in extended range forecasting. Ann Geophysicae 9: 242-251

Epstein ES (1969) Stochastic dynamic prediction. Tellus 21: 739-759

Fennessy MJ, Shukla J (1991) Comparison of the impact of the 1982/83 and 1986/87 Pacific SST anomalies on time-mean predictions of atmospheric circulation. J of Climate 4: 407-423

Ferranti L, Palmer TN, Molteni F, Klinker E (1990) Tropical-extratropical interaction associated with the 30-60 day oscillation and its impact on medium and extended range prediction. J Atmos Sci 47: 2177-2199

Hoffmann RN, Kalnay E (1983) Lagged average forecasting, an alternative to Monte Carlo forecasting. Tellus 35A: 100-118

Hollingsworth A, Cubasch U, Tibaldi S, Brankovic C, Palmer TN, Campbell L (1987) Mid-latitude atmospheric predictions on time scales of 10-30 days. In: Cattle H (ed) Atmospheric and oceanic variability. Royal Meteorological Society, London, p 117-151

Horel JD, Roads JO (1988) Sensitivity of regional predictability to flow characteristics. J Geophys Res 93: 11005-11014.

Johansson A, Saha S (1989) Simulation of systematic error effects and their reduction in a simple model of the atmosphere. Mon Wea Rev 117: 1658-1675.

Kalnay E (1990) An alternative Lagged Average Forecasting scheme. Research Activities in atmospheric and oceanic modelling, report n°14, WMO/TD-No.396: 6.22

Kalnay E, Livezey R (1985) Weather predictability beyond a week: an introductory review. In: Ghil M, Benzi R, Parisi G (eds) Turbulence and predictability in geophysical fluid dynamics and climate dynamics. North-Holland, p 311-346.

Kalnay E, Dalcher A (1987) Forecasting forecast skill. Mon Wea Rev 115, 349-356

Leith CE (1974) Theoretical skill of Monte Carlo forecasts. Mon Wea Rev 102: 409-418

Livezey RE (1990) Variability of skill of long-range forecasts and implications for their use and value. Bull Amer Meteorol Soc 71: 300-309

Lorenz EN (1982) Atmospheric predictability experiments with a large numerical model. Tellus 34: 505-513

Mansfield DA (1986) The skill of dynamical long-range forecasts, including the effect of sea surface temperature anomalies. Q J R Meteorol Soc 112: 1145-1176

McCalla C, Kalnay E (1988) Short and medium range forecast skill and the agreement between operational forecasts. Preprints of the 8th AMS Conference on Numerical Weather Prediction, Amer Meteorol Soc : 634-640

Michaud R (1990) Extended memory of the initial conditions in long-range forecasts of the January 1983 Atmospheric circulation. J Climate 3: 461-482

Miller AJ, Roads JO (1990) A simplified coupled model of extended-range predictability. J of Climate 3: 523-542

Milton SF, Richardson DS, Dickinson A (1991) Practical extended-range dynamical forecasting at UKMO. Extended abstracts submitted to the ICTP/WMO international technical conference on long-range weather forecasting research (Trieste, Italy, 8-12 april 1991), WMO/TD n° 395: 189-192

Miyakoda K, Sirutis J, Ploshay J (1986) One-month forecast experiments - without anomaly boundary forcings. Mon Wea Rev 114: 2363-2401

Miyakoda K, Sirutis J (1985) Extended range forecasting. Adv Geophys 28B: 55-85

Miyakoda K, Sirutis J (1990) Subgrid scale physics in 1-month forecasts. Part II: systematic error and blocking forecasts. Mon Wea Rev 118: 1065-1081

Mo KC, Zimmerman JR, Kalnay E, Kanamitsu M (1991) A general circulation model study of the 1988 US drought. Mon Wea Rev 119: 1512-1532

Mo KC, Kalnay E (1991) Impact of sea surface anomalies on the skill of monthly forecasts. Mon Wea Rev (in press)

Molteni F, Cubasch U, Tibaldi S (1987) 30- and 60-day forecast experiments with the ECMWF spectral models. Proceedings of the ECMWF workshop on predictability in the medium and extended range, 17-19 March 1986. ECMWF, Shinfield Park (UK), p 51-107

Molteni F, Tibaldi S (1990) Regimes in the wintertime circulation over northern extratropics. II: consequences for dynamical predictability. Quart J Roy Meteor Soc 116: 1263-1288

Molteni F, Mureau R, Palmer TN (1991) Linear instability analysis and its application to ensemble forecasting. Extended abstracts submitted to the ICTP/WMO international technical conference on long-range weather forecasting research (Trieste, Italy, 8-12 april 1991), WMO/TD n° 395: 195-200

Molteni F, Palmer TN (1991) A real-time scheme for the prediction of forecast skill. Mon Wea Rev 119: 1088-1097

Murphy AH (1988) Skill scores based on the mean square error and their relationships to the correlation coefficient. Mon Wea Rev 116: 2417-2424

Murphy AH, Winkler RL (1987) A general framework for forecast verification. Mon Wea Rev 115: 1330-1338

Murphy AH, Epstein ES (1989) Skill scores and correlation coefficients in model verification. Mon Wea Rev 117: 572-581

Murphy JM (1988) The impact of ensemble forecasts on predictability. Q J R Meteorol Soc 114: 463-493

Murphy JM, Palmer TN (1986) Experimental monthly long-range forecasts for the United Kingdom. Part II. A real time long range forecast by an ensemble of numerical integrations. Meteor Mag 115: 337-349

Murphy JM, Dickinson A (1989) Extended-range prediction experiments using an 11-level GCM. Meteorol Atmos Phys 40: 61-83

Murphy JM (1990) Assessment of the practical utility of extended range ensemble forecasts. Q J R Meteorol Soc 116: 89-125

O'Lenic EA, Livezey RE (1989) Relationships between systematic errors in Medium Range numerical forecasts and some of the principal modes of low-frequency variability of the Northern Hemisphere 700 mb circulation. Mon Wea Rev 117: 1262-1280

Owen JA, Palmer TN (1987) The impact of El Nino on an ensemble of extended range forecasts. Mon Wea Rev 115: 2103-2117

Palmer TN (1988) Medium and extended range predictability and stability of the Pacific/North American mode. Q J R Meteorol Soc 114: 691-714

Palmer TN, Tibaldi S (1988) On the prediction of forecast skill. Mon Wea Rev 116: 2453-2480

Palmer TN, Brankovic C, Molteni F, Tibaldi S (1990) Extended-range predictions with ECMWF models: Interannual variability in operational model integrations. Q J R Meteorol Soc 116: 799-834

Palmer TN, Brankovic C, Molteni F, Tibaldi S, Ferranti L, Hollingsworth A, Cubasch U, Klinker E (1990) The European Center for Medium-Range Weather Forecasts (ECMWF) Program on extended-range prediction. Bull Amer Meteorol Soc 71: 1317-1330

Roads (1989) Dynamical extended-range forecasts of the lower tropospheric thickness. Mon Wea Rev 117: 3-28

Royer JF, Déqué M (1990) Evaluation of one month forecast experiments with the French spectral model. Long-Range Forecasting Research Report n°13. WMO/TD n° 363: 161-166

Royer JF, Déqué M (1991) Improvement of long-range predictions by combination of model forecasts. Extended abstracts submitted to the ICTP/WMO international technical conference on long-range weather forecasting research (Trieste, Italy, 8-12 april 1991), WMO/TD n° 395: 235-240.

Saha S, Kalnay E, Kanamitsu M, Van den Dool HM (1990) DERF90 a long series of extended range (90-day) forecasts 3 may 1990 to 6 December 1990. Research Activities in atmospheric and oceanic modelling, report n°14, WMO/TD-No.396: 6.29-6.30

Saha S, van den Dool HM (1988) A measure of the practical limit of predictability. Mon Wea Rev 116: 2522-2526

Sausen R, Ponater M (1990) Reducing the initial drift of a GCM. Beitr Phys Atmosph 63: 15-24

Shukla J (1981) Dynamical predictability of monthly means. J Atmos Sci 38: 2547-2572

Shukla J (1984) Predictability of time averages: Part II: the influence of boundary forcings. In: Burridge DM, Källen E (eds) Problems and Prospects in Long and Medium Range Weather Forecasting. Springer-Verlag, Berlin Heidelberg New York Tokio, p 155-206

Shukla J (1985) Predictability. Adv Geophys 28B: 87-122

Shukla J, Gutzler DS (1983) Interannual variability and predictability of 500 mb geopotential heights over the Northern Hemisphere. Mon Wea Rev 111: 1273-1279

Shukla J, Fennessy MJ (1988) Prediction of time-mean atmospheric circulation and rainfall: influence of Pacific sea surface temperature anomaly. J Atmos Sci 45: 9-28

Sirutis J, Miyakoda K (1990) Subgrid scale physics in 1-month forecasts. Part I: experiment with four parameterization packages. Mon Wea Rev 118: 1043-1064

Smagorinsky J (1969) Problems and promises of deterministic extended range forecasting. Bull American Meteorol Soc 50: 286-311

Thompson PD (1957) Uncertainty of initial state as a factor in the predictability of large scale atmospheric flow patterns. Tellus 9: 275-295

Tibaldi S, Molteni F (1990) On the operational predictability of blocking. Tellus 42A: 343-365

Tibaldi S, Palmer TN, Brankovic C, Cubasch U (1990) Extended-range predictions with ECMWF models: Influence of horizontal resolution on systematic error and forecast skill. Q J R Meteorol Soc 116: 835-866

Tracton MS (1990) Predictability and its relationship to scale interaction processes in blocking. Mon Wea Rev 118: 1666-1695

Tracton MS, Mo K, Chen W, Kalnay E, Kistler R, White G (1989) Dynamical extended range forecasting (DERF) at the National Meteorological Center. Mon Wea Rev 117: 1604-1635

Tribbia JJ, Baumhefner DP (1988) Estimates of the predictability of low-frequency variability with a spectral general circulation model. J Atmos Sci 16: 2306-2317

Van den Dool HM (1985) Prediction of daily and time-averaged temperature for lead times of 1-30 days. Proc. Ninth Conf. on Probability and Statistics in Atmospheric Sciences, Virginia Beach, Amer Meteor Soc: 149-153

Van den Dool HM, Saha S (1990) Frequency dependence in forecast skill. Mon Wea Rev 118: 128-137

Walsh JE, Ross B (1988) Sensitivity of 30 day dynamical forecasts to continental snow cover. J Climate 1: 739-754

Yamada S, Maeda S, Kudo T, Iwasaki T, Tsuyuki T (1991) Dynamical one-month forecast experiments with the JMA global prediction model. J Meteor Soc Japan 69: 153-159

Zwiers FW (1987) A potential predictability study conducted with an atmospheric general circulation model. Mon Wea Rev 115: 2957-2974

OBSERVATIONAL AND MODELING STUDIES OF THE INFLUENCE OF SEA ICE ANOMALIES ON ATMOSPHERIC CIRCULATION

John E. Walsh
Department of Atmospheric Sciences
University of Illinois
105 S. Gregory Avenue
Urbana, Illinois 61801
U.S.A.

1. Background

For many decades the role of sea ice in climate variability has been the subject of considerable speculation. Scientific investigations based on data of varying quantity and quality have appeared rather regularly in the literature since the late nineteenth century. Much of the earlier work focussed on relationships between ice extent, storm tracks and surface temperatures. Apparent multi-year cycles in regional ice cover were also frequent topics in the earlier publications, although these cycles have generally failed the test of time. More recently, the emphasis has shifted to ice-ocean coupling and the role of sea ice in greenhouse warming. As the survey presented in Section 3 will indicate, the aggregate of sea ice research over the past few decades contains a "spectral gap" in the sense that there have been few applications to monthly and seasonal forecasting. However, ice-atmosphere interactions over shorter and longer timescales have been studied sufficiently to permit some meaningful inferences about the short-term climatic roles of sea ice.

The sea-ice data base increased rapidly during the post-World War II period with the advent of routine aircraft reconnaissance in the early 1950's and global satellite surveillance in the late 1960's and early 1970's. The useful global data on interannual variations of sea ice coverage now span approximately two decades, while regional datasets (Alaska, the North Atlantic) span four decades or longer. The scope of the analyses of ice-atmosphere interactions has also been expanded considerably in recent decades by the development of global climate models, many of which now include interactive oceans and sea ice. While these models permit controlled numerical experiments pertaining to air-sea-ice interactions, the success of fully coupled simulations has been limited thus far by shortcomings in the various component models.

In this review, we (1) compare the speculated and model-derived climatic roles of sea ice with the observational evidence and (2) assess the findings in the context of the prediction of interannual variations. Although the emphasis is on the prediction of atmospheric anomalies

NATO ASI Series, Vol. I 6
Prediction of Interannual Climate Variations
Edited by J. Shukla

over monthly and seasonal timescales, mention will also be made of the predictability of sea ice. The inclusion of sea ice predictability is motivated by the need for accurate boundary conditions in atmospheric models used for dynamic extended-range prediction.

2. Physical basis

Expectations concerning the climatic roles of sea ice are generally based on one or more of several properties of sea ice. Each of these properties can result in significant modifications of the fluxes of heat, salt and/or momentum into the lower atmosphere or upper ocean. Modifications of the air-sea heat exchange alter the diabatic heating of the lower atmosphere, thereby creating the potential for an influence on the atmospheric circulation.

The first and most visible effect of sea ice formation is an increase of the surface albedo from 0.05-0.15 over open water to 0.4-0.6 over snow-free ice or 0.6-0.9 over snow-covered ice. A second effect, attributable to the low thermal conductivity of sea ice, is a substantial reduction of the fluxes of sensible and latent heat at the surface. The reduction of these fluxes effectively insulates the atmosphere and ocean from each other in regions of thick and compact sea ice. A third effect of sea ice is the release and absorption of the latent heat associated with freezing and melting. Sea ice thus acts as a thermal reservoir that delays the seasonal temperature cycle. Finally, sea ice alters the salinity distribution and hence the ocean density stratification by the expulsion of brine during freezing and by the large-scale horizontal transport of the low-salinity ice. Because the dominant pattern of sea ice drift results in the transport of fresh water from the central Arctic to the Greenland and Iceland seas, sea ice transport is a major component of the freshwater budget of the North Atlantic region where much of the deep water of the world ocean is formed. Sea ice may therefore be an important factor in the variability of the global thermohaline circulation (Aagaard and Carmack, 1989).

Sea ice is a dynamic and complex mix of ice of different thicknesses. Sea ice moves in response to forces that include a spatially varying wind stress, water drag, Coriolis force, sea surface tilt and internal ice stress. Areas of thin ice and open water, continually created by deformation, are the sources of nearly all the new ice growth, salt rejection, and heat exchange with the atmosphere. Thick pressure ridges form due to convergence and shear in areas of thin ice. While the ice dynamics influence the ice thickness distribution, the thickness distribution in turn affects the ice dynamics by determining the ice strength. A sea ice cover is therefore a coupled dynamic-thermodynamic system.

A series of one-dimensional model experiments by Maykut (1978, 1982) showed the importance of the ice thickness distribution in determining the surface energy balance and the ocean-atmosphere coupling. Maykut's simulations were driven by seasonally varying climatological data for the central Arctic. The results showed that the winter-time heat input to the atmosphere is 1-2 orders of magnitude greater over thin ice (0-0.4 m) than over thick multiyear ice (2 m or more), although the results are sensitive to snow depth as well as ice thickness. The present distribution of ice thickness in the central Arctic appears to be such that the total heat input to the atmosphere from the small areal fractions (~0.05-0.10) of young ice and open water exceeds the fractions from the large areal fractions of thick ice. The timing of the summer melt process has also been found to be quite sensitive to the relatively small areas of open water, thin ice and melt ponds (Langleben, 1972). The thermodynamic role of sea ice is therefore considerably different from what it would be if sea ice were a uniform slab of ~3 m thickness.

3. Interactions between sea ice and the atmosphere

In order to place this survey into a framework, investigations of ice-atmosphere interactions over several timescales will be surveyed. First, we will review studies of interactions over the monthly and seasonal timescales that are directly relevant to interannual variability. Second, we will address interactions over the longer (multiyear to decadal) timescales of short-term climatic variability. Interactions over these longer timescales are appropriate here because (a) some of the conclusions obtained from investigations of these timescales have a direct bearing on monthly and seasonal prediction, and (b) the overwhelming majority of recent studies of sea ice-atmosphere interactions have addressed decadal-scale variability, including changes or trends associated with the greenhouse warming.

3.1 Monthly and seasonal timescales

3.1.1 Observational studies

Various observational studies have shown that variations of ice extent are associated with anomalies of air temperature locally (Fletcher, 1969; Niebauer, 1983; Fleming, 1987). However, statistical associations between ice cover and air temperature cannot be interpreted in terms of cause and effect. It is conceivable and even likely that other large-scale factors (e.g., abnormal poleward airflow) contribute to anomalies of both air temperature and ice coverage. In fact, if one is to search for causality in the cross-correlations derived from observational data, the most tenable interpretation is that ice anomalies are primarily responses to (not causes of) air temperature anomalies. Figure 1, for example, shows the lagged cross-correlations between sea ice and the 1000-500 mb thickness in the Davis Straight region of eastern Canada. As one would expect on physical grounds, ice and thickness are negatively correlated at small lags. In the 1-5 month range of lags, the cross-correlations are larger with the temperature leading the ice than with the temperature lagging the ice. This asymmetry has appeared in the cross-correlation functions between sea ice and other variables (Walsh and Johnson, 1979a) as

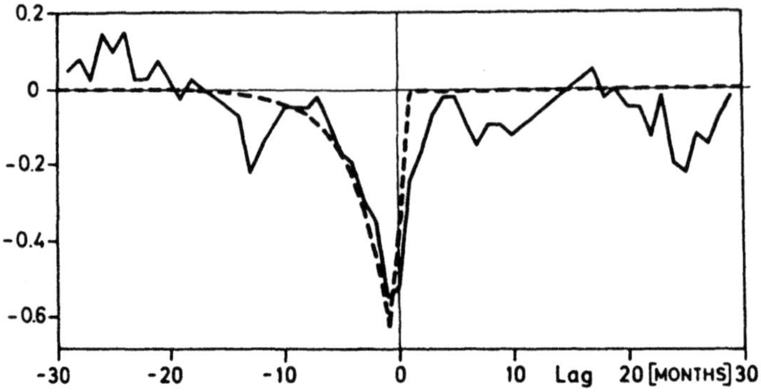

Figure 1. Cross-correlations between sea ice and the 500-1000 mb thickness in the Davis Strait (solid line), together with the theoretical curve (dashed line) for a first-order Markov model. Abscissa is lag (months) of the 1000-500 mb thickness (from Lemke et al., 1980, Fig. 22).

well as in correlations between sea surface temperatures and atmospheric variables (Davis, 1976). Such relationships illustrate the need for controlled model experiments (Section 3.1.2.) to address the extent to which sea ice actually affects the overlying atmosphere.

A more specific analysis of local associations was performed by Rogers (1978), who correlated northern Alaskan ice extent with monthly temperature and wind data from Barrow, Alaska. Sea ice generally recedes from the coast at Barrow, for 1-4 months during the summer. As indicated by the first line of Table 1, late-summer ice conditions are sensitive to early summer temperatures (of which the anomalies are largely advective in origin). Zakharov and Strokina (1978) have described similar relationships between sea ice and temperatures in the Soviet Arctic. Interestingly, Rogers found statistically significant correlations between September ice extent and air temperatures of <u>subsequent</u> months. These cross-correlations are larger than the autocorrelations (persistences) of the temperature anomalies at the same lags. Rogers' results suggest that at least a portion of the interannual variability of air temperature is attributable to the interannual variations of sea ice. The effect is local and seasonal, but it is statistically significant and strong enough to be of some use in monthly-to-seasonal forecasts of surface air temperatures for autumn.

TABLE 1. Correlation between the monthly temperature at Barrow and the distance from the northern Alaskan coast to the sea-ice edge (from Rogers, 1978).

Time of Ice Observation	Averaging Period for Temperatures	Correlation
Mid-September	June-September	0.82
Mid-September	September	0.77
Mid-September	October	0.61
Mid-September	November	0.44

With regard to ice-atmosphere interactions on the synoptic scale, one may cite a variety of studies of storm tracks in the context of regional sea ice variability. Studies of historical interest are those by Wiese (1924), Walker (1947) and Lamb (1972), all of whom show that storm tracks tend to shift northward and southward with the ice edge. For the Pacific sector, Overland and Pease (1982) present convincing evidence of associations between Bering Sea ice extent and wintertime cyclone trajectories. Cyclones are considerably more frequent in the western Bering Sea during light ice years and in the eastern Bering Sea during heavy ice years. Carleton (1981) finds that interannual fluctuations of cyclone activity and ice extent are also quite consistent in the vicinity of the Antarctic ice boundary. However, as in the case of local associations, the synoptic scale associations do not imply that either the cyclone trajectories or the sea ice boundary locations are causes of the other. Rather, natural or forced variations in the large-scale circulation may change the location of the subpolar baroclinic zones, thus influencing both the sea ice boundary and cyclone evolution. Because of the large natural variability of cyclone frequencies and trajectories, an unambiguous demonstration of sea-ice effects on cyclone variability is exceedingly difficult with observational data alone.

Possible forecasting applications of large-scale empirical relationships have been reported by Gongbing and Domrös (1987), who compiled monthly indices of Antarctic sea ice in various longitudinal sectors using data for 1973-1982. Correlations between these indices, various measures of the North Pacific subtropical high, and river runoff over China were claimed to be (a) statistically significant and (b) largest when the atmospheric/hydrologic variable lags the sea ice index by several months. While such findings are suggestive, they need to be confirmed with independent data and with Monte Carlo tests that take full account of the predictor-predictand selection procedure and of the spatial and temporal degrees of freedom in the data. The same statement applies to Lamb's (1982) correlations between 5-10 year averages of snow/ice indices and various large-scale atmospheric quantities.

The difficulty in distinguishing cause and effect in empirical relationships is one reason why the literature contains relatively few examples of statistical applications of sea ice data in long-range atmospheric forecasting. Another reason is that the sample of years is limited by the length of the period (≤ 20 years) of global sea ice datasets derived from satellites. Only recently (Parkinson, 1991) has it been possible to compile a meaningful depiction of interannual extremes of sea ice coverage based on a homogeneous dataset. Similarly, ice anomaly statistics (e.g., persistence) have only recently been evaluated from homogeneous data spanning more than 10-15 years (Chapman and Walsh, 1991). The latter study addressed the utility of ice anomaly persistence as a predictor of sea ice over monthly and seasonal timescales. The fractions of variance of ice coverage explained by anomaly persistence were found to decrease from ~.50 to ~.10 as the forecast range increases from several weeks to several months. These fractions are considerably larger than the corresponding values for atmospheric variables. Anomaly persistence from the winter months is generally larger than from the summer months. The results imply that the specification of temporally invariant anomalies of the sea ice boundary conditions should not contaminate a 30-day atmospheric model forecast as much as would a similar specification of continental snow cover, which generally varies more rapidly than sea ice.

3.1.2 Modeling studies

Perhaps because the horizontal scales of sea ice anomalies are smaller than those of sea surface temperature anomalies, there have been few studies of atmospheric model sensitivities to sea ice boundary conditions over 30-90 day timescales. The small number of such studies may, to at least some degree, be a consequence of the major research emphasis during the 1980's on tropical forcing of the global atmosphere. It can be argued theoretically that the basic state of the atmosphere is such that remote teleconnections are more likely to be excited by tropical surface anomalies than by high-latitude surface anomalies. The stronger static stability of the lower troposphere in high latitudes is often cited in these arguments.

Before surveying the general circulation model sensitivities to sea ice, we mention Adem's (1991) thermodynamic model, which has been used in 30-day forecast experiments. The initial conditions for these forecasts include positions of the snow and sea ice boundaries prescribed on the basis of observational data. The specification of an observed anomaly of snow cover has been found to improve the temperature forecasts in cases with large snow anomalies (Donn and Goldberg, 1982). However, there is no published evidence that forecasts by this model are improved by accurate sea ice boundary conditions.

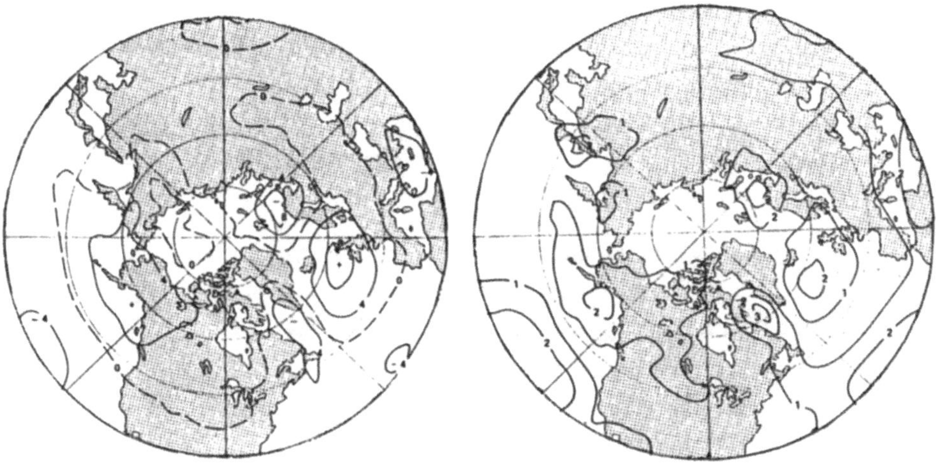

Figure 2. (a) Sea level pressure difference (mb) corresponding to pressure with minimum Arctic ice conditions minus pressure with maximum Arctic ice conditions, and (b) Ratio of absolute value of sea level pressure difference to standard deviation of sea level pressure (from Herman and Johnson, 1978, Figs. 3 and 4).

The most widely cited model experiment pertaining to sea ice sensitivities is Herman and Johnson's (1978) study of the response of an atmospheric general circulation model (GCM) to prescribed extremes of Arctic sea ice extent. In this experiment, a series of January simulations was performed with the GLA (Goddard Laboratory for Atmospheres) GCM. Two sets of prescribed ice extremes differing by ~10° latitude at all oceanic longitudes were used to generate subsets of model output. The prescribed ice extremes encompassed an area approximately twice as large as that encompassed by the corresponding observed extremes (Walsh and Johnson, 1979b; Parkinson, 1991). Sea surface temperatures were prescribed and held constant. Because the ice cover was a uniform slab of 3-m thickness, the ice contained no thin ice or leads. The two sets of model runs produced significant differences in sea level pressure (up to 8 mb), 700 mb geopotential height (up to 100 gpm) and tropospheric temperature (up to 8°C at low levels). Figure 2 shows the pressure response. While the largest differences occurred in the general vicinity of the ice margin, substantial differences in the pressure and height fields of the subtropical Atlantic and Pacific were also found. The model therefore suggests that a general advance or retreat of the ice edge by ~1000 km would have a significant impact over large portions of the Northern Hemisphere. It must be noted, however, that the representation of the ice as a 3-m slab without leads exaggerates the differences in surface forcing near the ice edge. (Early GCM experiments may have exaggerated the polar warming in response to doubled CO_2 for this same reason; Ingram et al. [1989] examine sea ice feedbacks in the context of doubled-CO_2 experiments performed with various GCMs.) Another caution concerning the Herman and Johnson experiment has emerged from data-based studies, which show that the most common sea ice anomaly patterns contain both positive and negative anomalies (e.g., Lemke et al., 1980). In other words, concurrent departures from normal ice extent tend to be of opposite sign in different longitudinal sectors. Hemispherically integrated anomalies therefore tend to be residuals rather than indicators of ice fluctuations in all longitudes.

Simmonds (1981) performed a similar wintertime experiment for the Southern Hemisphere by simulating the September atmospheric circulation (over 130 days) with (a) September ice extent in the Antarctic and (b) a reduced Antarctic ice extent corresponding to March observations. The reduction of ice extent shifted the largest sensible and latent heat fluxes poleward, thereby weakening the poleward temperature gradient and reducing the intensity of the westerly flow in the region south of 45°S (Fig. 3). The 500 mb height increased by up to 14 dm over the sea ice zone. In contrast to Herman and Johnson's (1978) results, sea level pressure increased in middle and high latitudes when ice cover was reduced. This increase of pressure is consistent with the September-to-March seasonal change observed in the Southern Ocean.

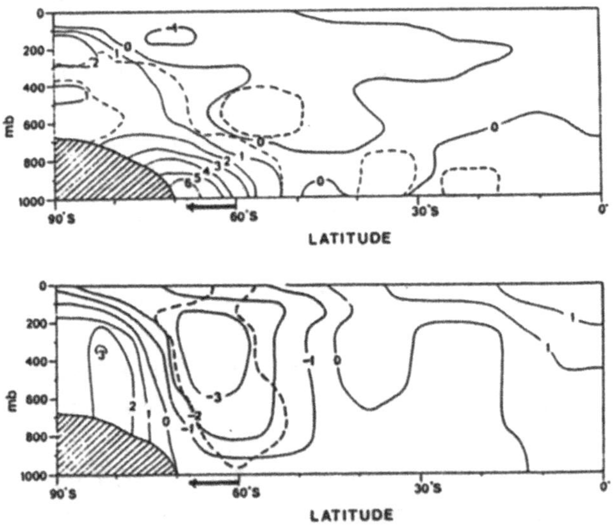

Figure 3. (a) Difference between zonally averaged temperature (°C) of the Antarctic ice anomaly (reduced ice) and control simulations. Arrow on abscissa indicates change in zonally averaged sea ice extent in the two runs. Regions of differences significant at the 90% confidence level are enclosed by dashed contours. (b) As in (a), but for zonal wind (m s^{-1}) (from Simmonds, 1981, Figs. 5-6).

Mitchell and Hills (1986) performed a more recent experiment in which all ice north of 66°S was removed in a series of winter (June-September) simulations. The large increase of sensible heat over the region of ice removal resulted in a large (> 10°C) but shallow cooling locally, a reduction of the westerly flow aloft, and a decrease of surface pressure near the new sea ice margin (Fig. 4). The decrease of pressure is consistent with Herman and Johnson's (1978) Arctic results, but not with Simmonds' (1981) Antarctic results. Because the treatment of sea ice in this model did not include leads, Mitchell and Hills (1986, p. 967) note that the "changes may be regarded as an upper bound to the errors due to the parameterizations" of the heat flux through the ice.

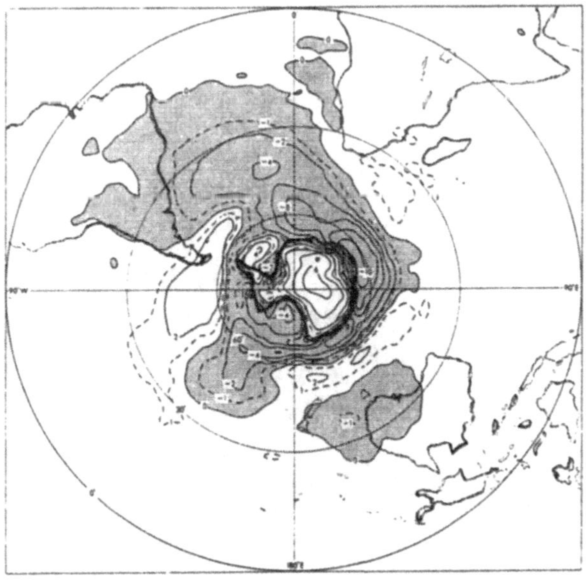

Figure 4. Changes in surface pressure due to removal of Antarctic sea ice equatorward of 66°S. Differences are averaged over July-September of three years. Areas of decrease are stippled (from Mitchell and Hills, 1986, Fig. 7a).

Walsh and Ross (1988) examined the sea ice sensitivities of 30-day forecasts produced by the R15 version of the NCAR Community Forecast Model (CFM). The CFM was run with initial atmospheric states and North Atlantic boundary conditions (sea ice and SST) derived from observational analyses for February 1 of two different years: 1984 (observed ice heavy) and 1982 (observed ice light). Figure 5 shows the differences in the boundary conditions of the two years. The simulated air temperatures were lower in the North Atlantic when more extensive ice was prescribed, and the differences of sea level pressure in the North Atlantic agreed well with those of Herman and Johnson (1978). However, the temperature response was confined to the lowest 150-200 mb of the experimental region (the North Atlantic). While the skill of the North Atlantic temperature forecasts was slightly higher when the boundary conditions were prescribed from observations, the pressure field showed no such

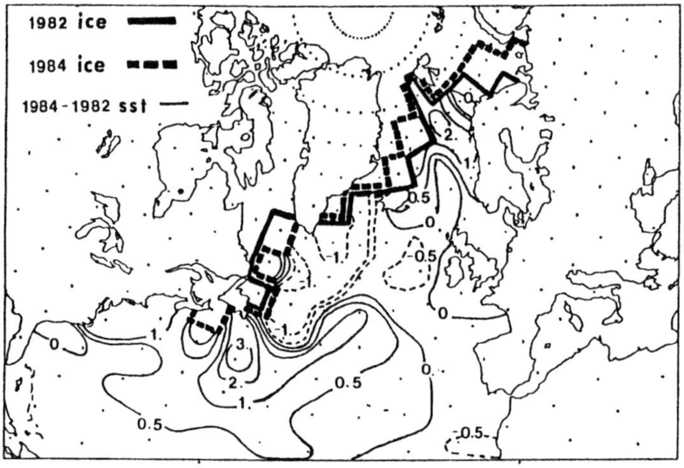

Figure 5. R15-truncated positions of the ice edge in 1982 (solid line) and 1984 (dashed line). Also shown are differences between North Atlantic sea surface temperatures of 1984 and 1982 (from Walsh and Ross, 1988, Fig. 1).

improvement. In order to demonstrate a positive impact of sea ice on the large-scale circulation, such experiments should include finer resolution (in order to resolve the boundary anomalies), a larger sample of cases, and multiple simulations of each case (e.g., lag average forecasts) in order to permit tests of the statistical significance of the sensitivities.

An increase of ice extent produced lower pressure in the Iceland - U.K. region of the North Atlantic in the experiments of Herman and Johnson (1978) as well as Walsh and Ross (1988). However, observational analyses show above-normal pressures over Iceland and Greenland when Icelandic sea ice is extensive (Kelly et al., 1987). This apparent discrepancy is likely attributable to the fact that southward airflow favors heavy ice near Iceland, thereby reinforcing the earlier contention that cause and effect are exceedingly difficult to untangle in observational analyses. Alternatively, Kelly et al. note that the ice anomaly prescribed by Herman and Johnson cannot be sustained near Iceland if the SLP response indeed results in intensification of the northward airflow in the Iceland - Greenland waters.

In the most recently published sea ice sensitivity experiment, Raymo et al. (1990) used the GISS II model to simulate the seasonal cycle of the response to a decrease of Arctic ice extent. Raymo et al. reduced the March ice extent to the current September ice extent, and they imposed areal reductions of the same magnitude in all other months. This procedure reduced the September ice extent to zero. The SST was constrained to be -1°C in all areas of ice removal. The major effects of the reduction of ice extent were a large seasonal warming in the circum-Arctic region (15-25°C in winter, 0-3°C in summer), northward migration of the Icelandic low pressure system, generally lower pressure in the Arctic, a strengthening of the Azores high (by 6-9 mb in winter), and increased precipitation and evaporation in high latitudes. The changes in the following quantities were statistically significant at the two-sigma level during the winter: hemispheric temperature, hemispheric snow cover, hemispheric planetary and surface albedos, precipitation and evaporation in the region poleward of 67°N, and regional sea-level pressures.

A noteworthy feature of the experiment by Raymo et al. (1990) is that the ice-covered regions contained a fraction (10%) of open water corresponding to leads. Leads were not included in the other experiments summarized above. For this season and because Raymo et al. did not permit high latitude ocean temperatures to increase above -1°C, the effect of the ice removal may have been underestimated. Nevertheless, as in the experiments on Herman and Johnson (1978) and Simmonds (1981), the differences in the prescribed extremes of sea ice exceed the observed interannual fluctuations by a factor of 2-3 and do not contain the longitudinal compensation that is observed on the interannual timescales.

3.2 Decadal timescales

The variations addressed in this section occur over multiyear and decadal timescales, which are generally longer than those of primary interest in the prediction of interannual variations. We include the decadal variations here because (a) such variations can contribute statistically to interannual departures from a climatological mean, and (b) some elements of the methodologies (e.g., modeling strategies) used to investigate decadal variability have a direct bearing on studies of the predictability of interannual variations.

3.2.1 Observational studies

As in the case of monthly and seasonal fluctuations, observational studies of decadal-scale fluctuations are constrained by the difficulties in unravelling the causal mechanisms underlying empirical associations. Lamb (1982), for example, lists high correlations between 5-year means of an index of Icelandic sea ice and (a) world temperature, (b) northern hemisphere temperature, and (c) temperatures and winds over the British Isles. Lamb also reports that 5-year means of winds over London and snow accumulation at the South Pole are correlated at +0.75. Even if one accepts the validity of the data, one cannot resolve the causal mechanisms underlying such correlations. It is likely that the large-scale circulation accounts for large fractions of the covariance between snow/ice and regional temperature, but the physical linkage between regional indices of ice/snow and hemispheric or global means (of temperature, for example) is speculative at best.

A more focussed approach to multiyear-decadal variations involving sea ice has been followed by Mysak et al. (1990), who show that sea ice anomalies (and associated anomalies of ocean salinity and temperature) propagate from the Greenland Sea to the Labrador Sea over timescales of 3-5 years. Mysak et al. then hypothesize a feedback to cyclone activity and high-latitude precipitation, which in turn leads to variations of freshwater runoff, Arctic sea ice and, eventually, export of freshwater as sea ice to the North Atlantic. The postulated timescale of the entire cycle is 15-20 years. While this interdecadal cycle is not inconsistent with the (sparse) hydrologic and atmospheric data of the twentieth century, several linkages are highly speculative and unsubstantiated as of yet. Nevertheless, such hypotheses provide a framework for observational analyses and modeling experiments in which sea ice is a key component.

3.2.2 Modeling studies

The literature of the past 20 years describes a series of GCM studies for which the relevant timescales are most appropriately regarded as decadal or longer. These studies are the so-called "ice-in vs. ice-out" experiments, in which the effect of the removal of all sea ice in one or both hemispheres is examined. Although there is no evidence that the Arctic has been ice-free at any time during the past several hundred thousand years, GCM simulations of the response to doubled CO_2 produce a sufficiently large polar warming that little sea ice remains in the summer hemisphere of the model in the $2 \times CO_2$ case.

Several GCMs of the 1970's were run with an ice-free Arctic (Fletcher et al., 1971; Newson, 1973; Warshaw and Rapp, 1973). Because similar experiments have been performed recently with more sophisticated GCMs, we summarize here the results of only the more recent experiments.

The effects of a complete removal of Antarctic sea ice in the Australian model have been described by Simmonds and Dix (1987). The "ice-out" simulation, which was run in a "perpetual July" mode, produced a zonally-averaged low-level warming of ~10°C in the region of ice removal, a cooling of ~2°C in much of the Antarctic stratosphere, and a weakening of the zonal westerlies by ~4 m s^{-1} through much of the atmosphere between 45°S and 70°S. These changes were statistically significant. In contrast to Mitchell and Hills' (1986) simulation with reduced Antarctic ice extent, sea level pressure did not decrease in the zone of ice removal; pressures in the subpolar South Atlantic actually increased by 4-5 mb. Simmonds and Dix contend that the model-to-model differences may be attributable to differences in the control runs, since a model's responses to anomalous forcing can depend strongly on its basic climatology (Palmer, 1985). The comparison of model responses to Antarctic sea ice removal nevertheless leads to the conclusion that "the nature of the local surface response to surface heating at mid to high latitudes is far from obvious" (Simmonds and Dix, 1987, p. 1401).

Two recent experiments pertaining to Arctic sea ice removal are described by Dümenil and Schröder (1989) and Royer et al. (1990). In the former study, which was motivated by possible CO_2-induced changes in sea ice, the German model was integrated at T21 resolution in a perpetual January mode with an ice-free Arctic Ocean. The model response included a large (5-10°C) temperature increase in high latitudes below approximately 800 mb and a reduction of 10-20 W m^{-2} in the outgoing longwave radiation to space from the Arctic. Surface pressure increased over a large portion of northern Asia. Low-level cloudiness increased, and the earth-atmosphere system as a whole cooled. This somewhat surprising negative feedback indicates that cloudiness and other elements of the climate system can complicate the response to an imposed anomaly in the forcing, even if the imposed anomaly is limited to high latitudes.

Royer et al. (1990) performed a similar experiment with a T42 version of the French spectral model. As in Dümenil and Schröder's (1989) experiment, the removal of the Arctic ice led to a warming of 5-15°C in the lower troposphere of the region poleward of 70°N. Sea level pressures over the Arctic Ocean and the Gulf of Alaska decreased by 5-8 mb, while comparable increases occurred over the North Atlantic Ocean and north-central Asia. Arctic and mid-latitude precipitation increased substantially (Fig. 6), in agreement with earlier studies of Arctic ice removal. Unlike Dümenil and Schröder, however, Royer et al. obtained a substantial reduction of cloudiness (from approximately 0.60 to 0.45) over the ice-free Arctic Ocean. The authors attribute this discrepancy to uncertainties in the parameterization of cloud processes and cumulus convection in high latitudes. The simulation of high-latitude cloudiness has been cited elsewhere (JOI, 1990) as a major obstacle to an improved understanding of the role of the Arctic in climate change.

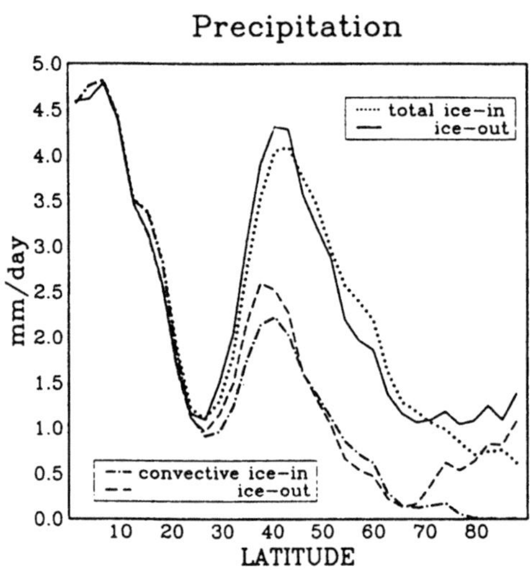

Figure 6. Zonally averaged precipitation (mm day^{-1}) for December - February in the Arctic "ice-in" and "ice-out" simulations of Royer et al. (1990). Upper curves represent total precipitation, lower curves, the part of precipitation due to convection (from Royer et al., 1989, Fig. 7).

An additional caution concerning the GCM experiments surveyed here pertains to leads. As noted in Section 2, leads profoundly affect the surface thermal regime in regions of sea ice. Most GCMs, including those discussed in this subsection, treat sea ice as a uniform slab without leads. Simmonds and Budd (1990) have recently examined the sensitivity of the Australian GCM (T21 resolution) to leads. In a "perpetual July" simulation, the specification of nonzero lead fractions (0.50 in the Antarctic, 0.05 in the Arctic) produced a warming of ~5°C over the Antarctic sea ice, a reduction of zonal winds over the Southern Ocean, and changes of 4-7 mb in the sea level pressure over large regions of the Southern Ocean and Antarctica (Fig. 7). The sensible heat fluxes at the surface changed by 100-200 W m^{-2} in southern high latitudes.

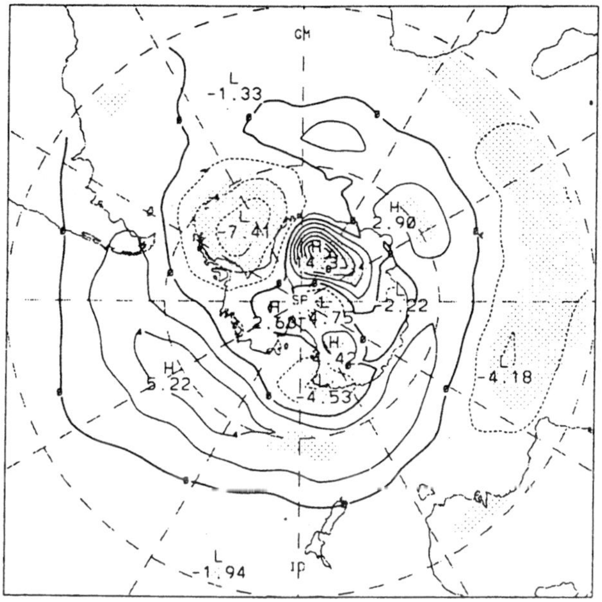

Figure 7. Difference between the mean July sea-level pressure of the "ice leads" and "control" simulations. Contour interval is 2 mb. Negative contours (i.e., lower pressure in the simulation with leads) are dashed, and regions of differences significant at 95% confidence level are stippled (from Simmonds and Budd, 1990, Fig. 3).

The lead-induced changes found by Simmonds and Budd are comparable to the changes obtained in the ice-removal experiments surveyed earlier. The minimum lead fraction is also a high-leverage parameter in energy balance models (Harvey, 1988a; Ledley, 1988). Improved parameterization of leads clearly merits high priority in the modeling atmosphere-ice interactions, not only in simulations of climate change but also in model simulations pertaining to interannual variability.

4. Conclusion

The preceding survey of sea ice in the context of short-term climate variability leads to several conclusions:

(a) Model studies and the limited observational analyses indicate that sea ice variations have large local impacts on the lower atmosphere. These impacts are large enough to be useful in local forecasting at the 30-90 day range.

(b) Observational studies are severely constrained by the fact that atmospheric forcing plays a major role in the development of sea ice anomalies. Statistical associations are of very limited use in the diagnosis of the climatic role of sea ice.

(c) Global climate models show significant sensitivities to large changes of sea ice coverage, e.g., the halving or complete removal of hemispheric ice coverage. While the responses of different models are similar in many respects, there are inconsistencies among the models in the changes of some fundamental variables (e.g., Arctic cloudiness, Antarctic sea level pressure).

(d) There has yet to be a convincing demonstration that interannual variations of sea ice produce a sufficiently large positive impact to be considered useful in 30-90 day forecasts produced by atmospheric models.

The impression that emerges from the aggregate of the studies surveyed here is that the findings to data suggest but do not unambiguously demonstrate the short-term climatic importance of sea ice. On the one hand, the local impacts of sea ice are large; climate models project the strongest greenhouse impacts in regions of sea ice; and energy balance models indicate that the global sea ice feedback is much stronger than the feedback from continental snow cover (Harvey, 1988b). The effects of sea ice on ocean salinity have potentially global impacts through the thermohaline circulation. On the other hand, global models have not been able to provide meaningful quantifications of these effects because of a combination of (i) inadequacies in the resolution and high-latitude parameterizations of the GCMs of the 1980's, (ii) limitations on atmosphere-ice-ocean model coupling imposed by the errors and biases in the component models, and (iii) the relatively limited attention given to high-latitude problems thus far by the global modeling community.

The results described in Section 3 point to several immediate priorities for improved modeling of the effects of sea ice on short-term climate variability. Among these priorities are the inclusion of leads and improved formulations of high-latitude cloudiness in GCMs. In view of the potentially important climatic role of sea ice albedo, there is a need for more accurate parameterization of the albedo as a function of the state of the sea ice surface (snow-covered, melting, ponded). Testing of formulations of sea ice properties and processes should proceed in parallel with more systematic comparisons of simulated and observed fields of the key atmospheric variables (pressure, temperature, precipitation, cloudiness), which must be simulated accurately if atmosphere-ocean-ice models are to be coupled successfully. One may argue that our understanding of the role of sea ice in interannual and climate variability will increase only with progress in the modeling of the atmosphere, ocean and sea ice as a coupled system.

Acknowledgements

The preparation of this paper was supported by the National Science Foundation, Climate Dynamics Program through Grant ATM-8912911 and by NASA's Interdisciplinary Research Program through Grant IDP-88-009. We thank Norene McGhiey for typing the manuscript.

References

Aagaard K, Carmack EC (1989) The role of sea ice and other fresh waters in the Arctic circulation. J Geophys Res 94:14485-14898
Adem J (1991) Review of the development and applications of the Adem thermodynamic climate model. Clim Dynamics 5:145-160
Carleton AM (1981) Ice-ocean-atmosphere interactions at high southern latitudes in winter from satellite observations. Austr. Meteorol Mag 29:183-195

Chapman WL, Walsh JE (1991) Long-range prediction of regional sea ice anomalies in the Arctic. Weather and Forecasting 6:271-288
Davis RE (1976) Predictability of sea surface temperature and sea level pressure anomalies over the North Pacific Ocean. J Phys Oceanogr 6:249-266
Dümenil L, Schröder S (1989) The impact of an ice-free Arctic Ocean on the general circulation in winter. Research Activities in Atmospheric and Oceanic Modelling (G Boer, Ed), WMO/TD-No 332:7.42-7.43
Donn W, Goldberg R (1982) Thermodynamic model: Sensitivity to input data. In: Proc Sixth Annual Climate Diagnostics Workshop, Washington, DC, NOAA, 311-315
Fleming GH (1987) Predictability of ice concentration in the high-latitude North Atlantic from statistical analysis of SST and ice concentration data. M Sc Thesis, Naval Postgraduate School, Monterey CA, 124 pp
Fletcher JO (1969) Ice extent on the Southern Ocean and its relation to world climate. Rand Corp Memo RM-5793-NSF, 111 pp
Fletcher JO, Mintz Y, Fox T (1979) Numerical simulations of the influence of Arctic sea ice on climate. WMO Tech Note 129:181-218
Gongbing P, Domrös M (1987) Statistical studies of the atmospheric circulation of the Northern Hemisphere, hydroclimatic regimes in China and Antarctic ice-snow cover. Large Scale Effects of Seasonal Snow Cover (Goodison BE, Barry RG, Dozier J, Eds), IAHS Press, Wallingford UK, 61-72
Harvey LDD (1988a) Development of a sea ice model for use in zonally averaged energy balance climate models. J Climate 1:1221-1238
Harvey LDD (1988b) On the role of high latitude ice, snow, and vegetation feedbacks in the climatic response to external forcing changes. Clim Change 13:191-224
Herman GF, Johnson WT (1978) The sensitivity of the general circulation to Arctic sea-ice boundaries. Mon Weather Rev 106:1649-1664
JOI (1990) Arctic System Science: Ocean-Atmosphere-Ice Interactions. Joint Oceanographic Institutions, Inc, Washington DC, 132 pp
Ingram WJ, Wilson CA, Mitchell JFB (1989) Modeling climate change: An assessment of sea ice and surface albedo feedbacks. J Geophys Res 94:8609-8622
Kelly PM, Goodess CM, Cherry BSG (1987) The interpretation of the Icelandic sea ice record. J Geophys Res 92:10835-10843
Lamb HH (1982) The climate environment of the Arctic Ocean. In: Rey L (Ed) The Arctic Ocean: The Hydrographic Environment and the Fate of Pollutants. MacMillan, London 135-161
Lamb HH (1972) Climate: Past, Present and Future. Volume I: Fundamentals and Climate Now. Methuen, London, 835 pp.
Langleben MP (1972) The decay of an annual cover of sea ice. J Glaciol 11:337-344
Ledley, TS (1988) A coupled energy balance climate-sea ice model: Impact of sea ice and leads on climate. J. Geophys Res 93:15919-15932
Lemke P, Trinkl EW, Hasselmann K (1980) Stochastic dynamic analysis of polar sea-ice variability. J Phys Oceanogr 10:2100-2120

Maykut GA (1982) Large-scale heat exchange and ice production in the central Arctic. J Geophys Res 87:7971-7984

Maykut GA (1978) Energy exchange over young ice in the central Arctic. J Geophys Res 83:3646-3658

Mitchell JFB, Hills TS (1986) Sea-ice and the antarctic winter circulation: A numerical experiment. Quart J Roy Meteor Soc 112:953-969

Mysak LA, Manak DK, Marsden RF (1990) Sea ice anomalies observed in the Greenland and Labrador Seas during 1901-1984 and their relation to an interdecadal Arctic climate cycle. Clim Dynamics 5:111-133

Newson, ML (1973) Response of a general circulation model of the atmosphere to removal of the Arctic ice-cap. Nature 241:39-40

Niebauer HJ (1983) Multiyear sea-ice variability in the eastern Bering Sea: An update. J Geophys Res 88:2733-2742

Overland JE, Pease CH (1982) Cyclone climatology of the Bering Sea and its relation to sea-ice extent. Mon Weather Rev 110:5-13

Palmer TN (1985) Response of the UK Meteorological Office General Circulation Model to sea surface temperature anomalies in the tropical Pacific Ocean. In: Nihoul JCJ (Ed), Coupled Ocean-Atmosphere Models, Elsevier Oceanogr Series 40:83-107

Parkinson CL (1991) Interannual variability of the spatial distribution of sea ice in the north polar region. J Geophys Res 96:4791-4802

Raymo ME, Rind D, Ruddiman WF (1990) Climatic effects of reduced Arctic sea ice limits in the GISS II general circulation model. Paleoceanography 5:367-382

Rogers JC (1978) Meteorological factors affecting the interannual variability of summertime ice extent in the Beaufort Sea. Mon Weather Rev 106:890-897

Royer JF, Planton S, Deque M (1990) A sensitivity experiment for the removal of Arctic sea ice with the French spectral general circulation model. Clim Dynamics 5:1-17

Simmonds I (1981) The effect of sea-ice on a general circulation model of the southern hemisphere. In: Allison I (Ed) Sea Level, Ice and Climate Change, Int Assoc Hydrol Sci, IAHS Publ No 131:193-206

Simmonds I, Budd WF (1990) A simple parameterization of leads in a general circulation model, and the sensitivity of climate to change in Antarctic ice concentration. Ann Glaciol 14:266-269

Simmonds I, Dix M (1987) Comment on 'Sea ice and the antarctic winter circulation: A numerical experiment' by J.F.B. Mitchell and T.S. Hills. Quart J Roy Meteor Soc 113:1396-1403

Walker G (1947) Arctic conditions and world weather. Quart J Roy Meteor Soc 73:226-256

Walsh JE, Johnson CM (1979a) Interannual atmospheric variability and associated fluctuations in arctic sea-ice extent. J Geophys Res 84:6915-6928

Walsh JE, Johnson CM (1979b) An analysis of arctic sea ice fluctuations. J Phys Oceanogr 9:580-591

Walsh JE, Ross B (1988) North Atlantic sea ice sensitivities of monthly dynamical forecasts. Preprints, Second Conf on Polar Meteorology and Oceanography, Amer Meteor Soc, Boston, MA 32-35

Warshaw M, Rapp RR (1973) An experiment on the sensitivity of a global circulation model. J Appl Meteor 12:43-49

Wiese W (1924) Polareis und atmosphärische Schwankungen. Geog Annal 6: 273-299

Zakharov VF, Strokina LA (1978) Recent variations in extent of Arctic Ocean ice cover. Meteorol Gidrol 7(1978):35-43.

OBSERVATIONAL AND MODELING STUDIES OF THE INFLUENCE OF SNOW ANOMALIES ON THE ATMOSPHERIC CIRCULATION

John E. Walsh
Department of Atmospheric Sciences
University of Illinois
105 S. Gregory Avenue
Urbana, Illinois 61801
U.S.A.

1. Introduction

Snow cover has long been regarded as an indicator and as a possible agent of climate variability over a variety of timescales. Over periods of several days to several weeks, the largest changes in the earth's surface properties result from variations in continental snow cover and snow-covered sea ice. As will be shown in Section 2, snow cover clearly influences the local values of near-surface atmospheric variables over these timescales. Over timescales of centuries and longer, the advance and retreat of the ice sheets depend ultimately on changes in the rates at which snow accumulates and ablates over the continents. However, on the interannual timescales relevant to monthly and seasonal forecasting, the effects of snow cover are relatively poorly understood. While there is indeed suggestive evidence that monthly and seasonal anomalies of snow affect the atmospheric circulation, the short-term climatic role of snow cover has not been clearly defined, nor has its utility in monthly and seasonal forecasting been established. This paper contains a survey of the research to date on snow cover's effects on the atmosphere over the three timescales: short-term (several days), interannual (several weeks to several seasons), and decadal (several years to several decades). The ambiguities as well as the consistencies of the various findings will be noted throughout the survey, which includes observational as well as modeling studies.

2. Physical background

The major difficulty in quantifying the climatic role of snow cover is that the distribution of snow is primarily a consequence of the large-scale pattern of atmospheric circulation, which determines the broad features of the distributions of temperature and precipitation over land. Thus, even in the absence of any causal role of snow cover, there can be large statistical correlations between anomalies of snow cover and atmospheric circulation. However, the pattern of the atmospheric circulation itself is determined by the

NATO ASI Series, Vol. I 6
Prediction of Interannual Climate Variations
Edited by J. Shukla

distribution of diabatic heating--radiation, conduction, and latent heating. By modifying the exchange of energy between the surface and the atmosphere, snow cover alters the distribution of diabatic heating in the atmosphere. For example, the albedo of fresh snow is 0.80-0.85 in visible wavelengths, whereas the albedo of bare land or ice-free ocean is typically between 0.05 and 0.30. Snow cover can therefore reduce the solar energy available to the surface and lower atmosphere by 50% or even more, depending on the age and depth of the snow, the vegetative cover and cloudiness. If this energy deficit is distributed through the lowest 2 km of the atmosphere, it can be equivalent to a cooling of 3-7°C in middle latitudes under clear skies during March (Namias, 1962). Snow cover is also an effective insulator of the underlying surface and an effective radiator of infrared energy. Finally, melting snow represents an effective sink of (latent) heat for the atmosphere and an effective source of moisture for the soil. The subsequent evaporation of this moisture may prolong the tendency for snow to delay the sensible heating of the soil.

A striking feature of the present-day distribution of snow cover over land is the virtual absence of seasonal snow in the Southern Hemisphere, where the only large area of land at latitudes in which snow can easily accumulate is the glaciated Antarctic continent. The following discussion of snow cover in the context of interannual variability is therefore limited almost exclusively to the Northern Hemisphere.

3. Applications of snow cover to atmospheric prediction

3.1 Short timescales (several days)

While timescales of several days are shorter than those of interest in monthly and seasonal forecasting, short-term effects are noteworthy because (1) they permit diagnostic assessments of the physical processes through which snow affects the atmosphere, and (2) their magnitude, duration and spatial scale determine whether monthly or seasonal atmospheric fields will be impacted by anomalies of snow cover.

The <u>local</u> impacts of snow on surface temperature can indeed be substantial over timescales of several days. Analysis of rates at which relatively warm moist air is cooled as it flows over snow indicate that the loss of heat by conduction to the surface can reduce the surface air temperature by 4-5°C per day (Treidl, 1970). Petersen and Hoke (1989) present a more recent example in which the accurate specification of snow cover improved a regional numerical model's 48-hour forecast of surface temperature (by 8-9°C) as well as the corresponding forecast of precipitation (rain instead of snow). Pan et al. (1989) report that improvements attributable to more accurate snow specifications have also been found in the global spectral model used by the U.S. National Meteorological Center. Ross and Walsh (1986) examined a 100-case sample of simple model (barotropic; analog) forecasts to show that extensive snow cover (and, by implication, enhanced coastal baoclinicity) contribute to stronger cyclone intensification and/or to storm motion parallel to the snow margin in eastern North America. Namias (1962) obtained similar results for selected cases from the winter of 1960-61.

The radiative impact of the surface albedo enhancement by snow can depress daytime surface temperatures by 5-10°C during spring, as shown by Dewey's (1977) diagnosis of errors in statistical forecasts that ignored snow cover (Fig. 1). If this approach is extended to monthly temperature specifications based on upper-air geopotential, the corresponding errors are found to be 5-7°C equatorward of normal snow boundary during months of extreme snow cover (Namias, 1985). When all winter months over ~30 years are included, the errors decrease to 1-3°C (Walsh et al., 1985). These impacts of snow are generally larger in spring, when insolation is stronger. However, in cases of positive anomalies of snow cover, this suppression of temperature is generally confined to the lowest 100-200 mb (Namias, 1985). Lamb (1955) showed that the 1000-500 mb thickness decreases as a layer of air passes over a large area of snow cover, implying that snow cover can contribute to the maintenance of a trough of cold air, which in turn helps to maintain the snow cover.

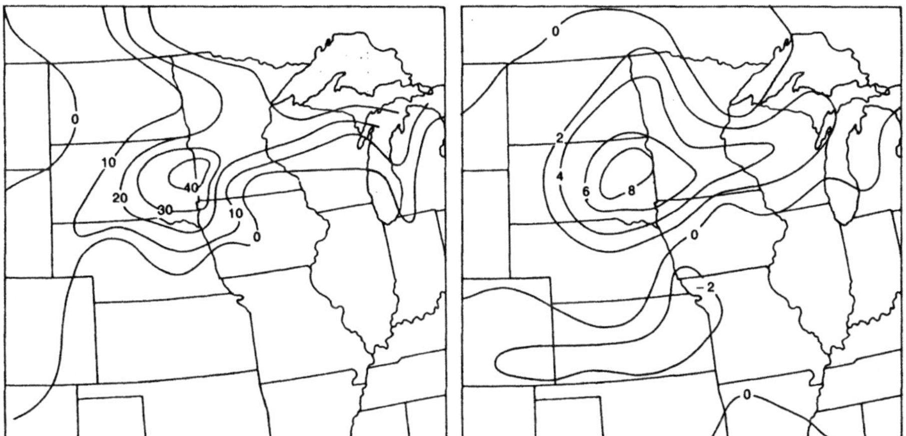

Figure 1. (a) Snow depth (cm) at 1200 UTC on 5 March 1977, and (b) 5-day average of the error (°C) of the daily maximum temperature forecasts for 5-9 March 1977 derived from National Meteorological Center's Model Output Statistics (from Dewey, 1977, Figs. 2 and 3).

Because the wintertime troposphere over land areas is characterized by relatively strong static stability even without a suppression of the near-surface temperature, it is unlikely that conditions are favorable for the vertical propagation of the thermal anomalies produced by snow cover. Nevertheless, the studies summarized above indicate that the magnitude of the effect of snow on surface temperatures and on short-term temperature forecasts is sufficiently large, at least locally and regionally, to merit consideration in numerical weather prediction. Whether or not this conclusion extends to monthly and seasonal forecasts depends on the persistence and/or predictability of anomalous snow cover.

3.2 Interannual timescales (several weeks to several seasons)

Because the interannual timescales represent the focus of this workshop, we include separate subsections on the relevant observational and modeling studies. The strategy is to survey first the evidence and speculation that have emerged from the analyses of observational data, and then to assess the extent to which model studies have supported, contradicted or extended the observational findings. Since causal mechanisms are often difficult to untangle in observational studies, numerical models offer the potential not only for useful predictions but also for controlled experimentation to increase our scientific understanding of the effects of snow.

3.2.1 Observational studies

Some of the earliest work on the role of snow cover in monthly and seasonal atmospheric variability was directed at the predictability of the Asian monsoon. Blanford (1884), Walker (1910) and others postulated a link between Himalayan winter snow cover and the strength of the Indian summer monsoon by reasoning that extensive snow cover could retard the heating of the Asian landmass. This link has been substantiated in recent years with the aid of satellite-derived measurements of snow cover (Hahn and Shukla, 1976; Dey et al., 1984; Dickson, 1984; Bhanu Kumar, 1988). On the basis of data for the 1967-1980 period, linear correlations between December-March snow extent in the Himalayas and June-September monsoon rainfall were found to be approximately -0.6 (Dey et al., 1984; Dickson, 1984). Dey et al. (1984) reported that rainfall in north-central India correlated more highly with the antecedent Himalayan snow cover than did rainfall in southern India. Correlations of Himalayan snow with the monsoon onset date over Kerala, the advance period, and the withdrawal period were found by Dey et al. to be of a similar magnitude (approximately 0.6).

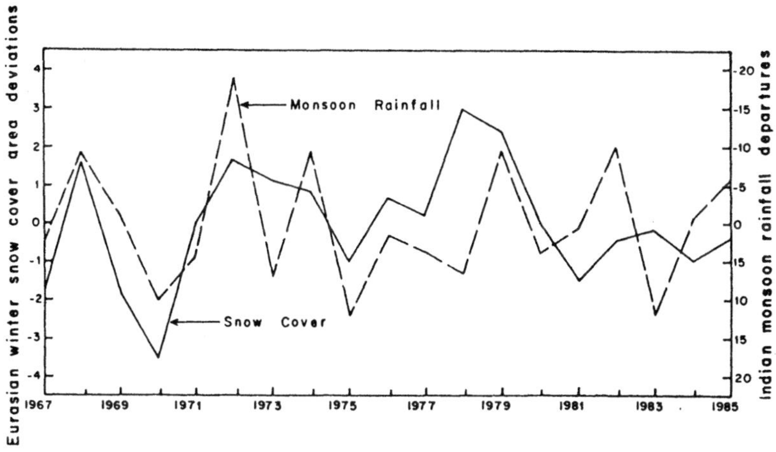

Figure 2. Variations of Eurasian winter snow cover area and the corresponding summer monsoon rainfall over India (from Bhanu Kumar, 1987, Fig. 4).

While the monsoon-snow correlations seem to provide the basis for useful monsoon predictions at ranges of several months, there are several important caveats. First, snow cover over the Himalayan region was not mapped consistently during the first part (1967-1972) of the satellite record (Ropelewski et al., 1984). Second, as shown in Fig. 2 (from Bhanu Kumar, 1987), the agreement between the interannual fluctuations of snowfall and monsoon variables has been noticeably poorer since 1980. Inclusion of data from 1981-1985 lowered the snow-rainfall correlations from -0.60 to -0.38 ($r^2 \approx 0.14$). Third, Indian monsoon rainfall appears to correlate as highly with snow cover over the remainder of Eurasia as it does with Himalayan snow cover (Dickson, 1984). The physical linkage between snowfall and monsoon rainfall may therefore be more complicated than implied by the proposed effect of snow on the timing of the heating of the Himalayan and Tibetan region.

Other observationally-derived linkages between snow cover and the atmospheric circulation over monthly timescales have been addressed by Afanas'eva et al. (1979), who examined the position of the "Planetary Upper-air Frontal Zone" (PUFZ) over Eurasia during the autumn and spring. Time variations of the PUFZ and the snow boundary correspond closely ($r = 0.78$). However, as noted in Section 2, such correlations are at least partially attributable to the fact that the position of major upper-air features (e.g., the jet stream) is a primary determinant of the position of the snow boundary. Interannual variations of snow cover over Eurasia were also examined by Toomig (1981), who found that the annual value of the absorption of solar radiation at Soviet stations is a strong function of the springtime absorption, which in turn depends on the albedo during spring. Toomig also reported modest correlations (r $\approx$ 0.45) between the surface albedo of early spring and the surface air temperatures through July at several stations. A similar lag relationship was suggested by Lamb (1972), who argued that the exceptionally cold summer of 1968 in northeastern Europe and northwestern Asia was attributable in part to the previous winter's extensive snow cover. Because the snow did not retreat until June and July, the ground remained cold and wet, thereby favoring the persistence of a trough. The attendant cloudiness and frequent cyclones produced above-normal rain and occasional snowfall throughout the summer over the northern USSR. Wahl (1968) suggested that persistent snow may have played a similar role in the unusually cold summers of the 1830's in northern North America.

Lag relationships between snow and air temperature were also evaluated by Foster et al. (1983), who found a surprisingly high correlation (r = -0.72) between Eurasian snow cover of October-November and the air temperature of the subsequent winter at several stations in the central USSR. Foster et al. suggest that when the Asian anticyclone develops early, which is more likely when there is extensive snow cover in the autumn, the anticyclone is stronger and colder than normal. Since this anticyclone dominates the continental interior with clear skies and light winds during winter, much more so than does its North American counterpart, large parts of Eurasia are colder than normal when the anticyclone is well developed.

Feedbacks involving snow cover have been suggested by Namias (1962, 1981), who has argued that extensive snow cover over eastern North America should enhance the coastal baroclinicity and favor strong cyclones along the East Coast. These systems, in turn, reinforce the cold trough and contribute to its persistence over periods of several weeks or longer, as in the winters of 1960-61 and 1976-77. Dickson and Namias (1976) have attempted to link extensive snow/extreme cold over eastern North America with anomalies in the winter climatic regime of Greenland and northwestern Europe. According to the authors' hypothesis, the southward displacement and intensification of cyclones on the East Coast of the United States lead to a reduction of cyclone activity over the Iceland - Greenland region. This weakening of the Icelandic low reduces the advection of mild air into the northeastern North Atlantic and northern Europe, thereby increasing the frequency of polar outbreaks over the European land areas. The frequent recurrence of this abnormal winter regime during the 1960's may have contributed to the relatively extensive European snow cover during that decade. However, the role of snow cover in modifying the atmospheric circulation in such regimes remains speculative because other hypotheses (involving SST anomalies, tropical forcing, etc) can be formulated. There is a need for controlled model experiments to explore the physical basis of seasonal anomalies in these specific cases, as has been done for anomalies such as the North American drought of 1988 and the ENSO event of 1982-83.

Finally, several studies have addressed the monthly and seasonal predictability of continental-scale anomalies of snow cover itself. Wiesnet and Matson (1976) showed that the January - March anomalies of Eurasian and Northern Hemisphere snow cover can be forecast with some skill by using the corresponding anomalies of the preceding December. However, Wiesnet and Matson's results were based on only nine years of data. More recently, Iwasaki (1991) confirmed the tendency for large-scale anomalies of snow cover to persist from December through February, but he found no indication of anomaly persistence from February to March in the longer data record. Iwasaki also reported an apparent lag relationship, in which winters with extensive snow cover over eastern Eurasia tend to be followed by winters with extensive snow cover over North America. Although this lag relationship is statistically significant ($r = 0.52$), at least three counter-examples occurred during the 1980's.

3.2.2 Modeling studies

As the preceding subsection indicates, various roles have been attributed to snow cover in the evolution and persistence of monthly and seasonal atmospheric anomalies. Most of these roles involve feedbacks, which are notoriously difficult to unravel from observational data. Consequently, model experiments provide a potentially useful vehicle for diagnosing the effects of snow cover. In this survey of modeling relevant to monthly and seasonal prediction, we proceed from studies based on relatively simple models to those based on state-of-the-art general circulation models. The mixed nature of the conclusions pertaining to snow cover as a predictive tool will become apparent as the survey proceeds.

Roads (1981) used a nonlinear primitive-equations model with an idealized land-sea distribution to obtain support for Namias' (1962) hypothesis that extensive snow cover over eastern North America generates anomalous offshore low pressure and inland high pressure. The eddy fluxes were found to be essential for forcing or dissipating the stationary eddy component of the atmospheric response to snow cover. A general conclusion of Roads merits consideration in the assessments of the other studies surveyed here: " . . . even in models which contain relatively few degrees of freedom in comparison to global general circulation models, it is difficult to understand why a particular [atmospheric] anomaly occurred in response to a change in the boundary conditions" (Roads, 1981, p. 7423).

Robock and Tauss (1986) addressed monthly and seasonal predictability by incorporating the forcing by snow cover into the simple, linear steady-state climate model of

Opsteegh and Mureau (1984). Results were obtained for October, January and April of the winters of 1976 through 1982. The pattern correlations of the simulated and observed fields were not improved by the addition of the observed snow cover as a forcing mechanism, suggesting that " . . . the effect of snow cover is much less than previously thought and/or the many other variables not parameterized in this model are much stronger factors than snow cover" (Robock and Tauss, 1986, p. 210). However, Robock and Tauss' results are not inconsistent with the findings of Roads, whose attempt to understand his primitive-equations model results as a linear response to diabatic heating was unsuccessful because of the importance of the anomalous eddy fluxes.

Walsh and Ross (1988) used the R15 version of the NCAR Community Forecast Model (CFM) in an experiment consisting of a pair of 30-day forecasts for each of ten cases from the January - February periods of 1977 through 1983. The two simulations of each case differed only in the extent of snow prescribed over eastern North America; observed snow cover in this region was above-normal in five of the cases and below-normal in the other five cases. Sea surface temperatures were prescribed climatologically. The root-mean-square errors of the forecasts of sea level pressure were not systematically different in the simulations corresponding to the observed snow cover of the initial state (Fig. 3). However, it should be noted that the observed snow cover varied considerably, even to the extent of a change in the sign of the anomaly, during the 30 days of almost every simulation. This rapid variability of snow cover indicates the need for either (a) specification of the time-dependence of the snow anomalies in experiments such as these, or (b) use of a larger pool of potential cases in order to identify 30-day winter periods in which the snow anomalies do not change substantially.

In a study of snow-induced effects in non-winter months, Yeh et al. (1983) used a simplified version of the GFDL general circulation model containing idealized geography and a limited computational domain. The model contained no diurnal cycle, and its cloudiness was prescribed to be zonally uniform and seasonally invariant. The complete removal of the snow cover in mid-March was found to reduce the water available to the soil through snowmelt, thus decreasing the soil moisture during the spring and summer in the region of snow removal. The drying of the soil resulted in an increase of surface temperature at high latitudes by 2-8°C for the subsequent 3-4 months. The temperature increase extended into the upper troposphere, thereby reducing the meridional temperature gradient and the zonal wind in high latitudes.

A similar conclusion about the snow-hydrology-temperature linkage was obtained from a more realistic GCM by Yasunari et al. (1989). In this experiment, the 5° × 4° version of the Japanese Meteorological Institute GCM was run for six months beginning March 1.

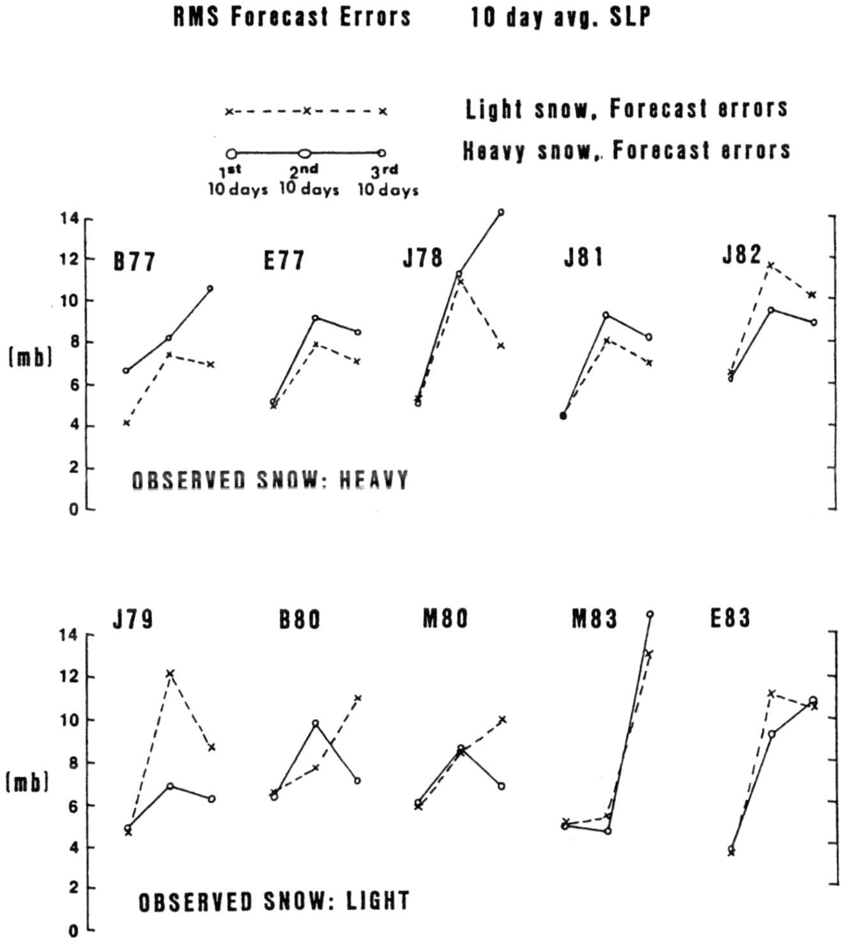

Figure 3. Root mean square errors of 30-day NCAR CFM forecasts of sea level pressure in the North Atlantic - North American region bounded by 20°N, 70°N, 7.5°W and 120°W. Errors are shown for each 10-day period of the forecasts made with boundary conditions of heavy snow (solid lines) and light snow (dashed lines). Calendar year (- 1900) of each case is indicated by final two digits of label (from Walsh and Ross, 1988, Fig. 6).

The experimental runs were identical to the control runs except for the addition of 5 cm (water equivalent) of snow in the snow-covered portion of the 30°-60°N zone of the Eurasian continent. The results (Fig. 4) contained evidence of both (1) an albedo feedback, which suppressed temperatures over lower latitudes (e.g., Tibetan Plateau) by 2-3°C during spring, and (2) a snow hydrology-temperature linkage, which suppressed temperatures over middle latitudes by 2-3°C during the summer months of June-August. During the summer, the anomalous Eurasian heat sink also appeared to induce a stationary Rossby wave pattern extending from eastern Asia to northern North America.

The most thorough investigation of spring-summer feedbacks involving snow cover is Barnett et al.'s (1989) study of Eurasian snow impacts on a low-resolution (T21) version of the ECMWF model. In Barnett et al.'s first experiment, snow extent corresponding to observed extremes was prescribed and interactions between snow and the surface hydrology were suppressed in order to isolate the albedo effect. The atmospheric response to the snow anomalies was local and confined primarily to air temperature and upper-air geopotential (but not sea level pressure). All significant signals vanished when the snow disappeared in the spring, and the albedo effect had no sustained impact on the development of the Asian monsoon.

In Barnett et al.'s second experiment, rates of snowfall over Eurasia were doubled and halved so that the subsequent melt and evaporation could induce changes in the regional hydrology. The two sets of simulations showed statistically significant differences extending through the following two seasons. The results derived from the heavier snow were characterized by significantly lower surface and tropospheric temperatures from May through July, higher pressures over Asia and lower pressures over North America (Fig. 5), weaker zonal winds over the Arabian Sea, weaker surface convergence over southern Asia, and a weaker monsoon over southeast Asia (Fig. 6). The Indian monsoon, however, was not substantially weaker in the "heavy snow" simulations, although this result may be partially attributable to the model resolution. The sea level pressure signal over Asia and North America (Fig. 5) is stronger in the model than in the real world; the exchange of mass between the two continents may have been exaggerated in the model because the SST distribution was prescribed climatologically in all the model runs.

In general, the physical mechanisms underlying Barnett et al.'s model response are similar to those of Yeh et al. (1983). The doubling of snowfall by Barnett et al. also produced a general weakening of the wind in the Southern Hemisphere and along the equator. While this response is similar to that which occurs prior to the warm phase of an ENSO event, the model with prescribed SST cannot sustain an ENSO event. In order to

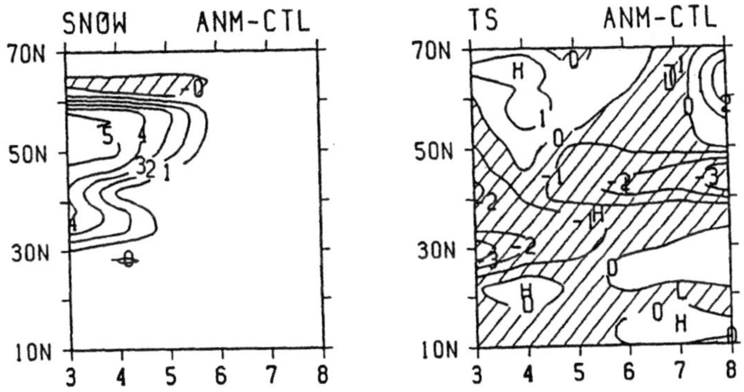

Figure 4. Latitude-time (month) sections of anomalies (HEAVY SNOW - CONTROL) of snowmass (cm liquid equivalent) and surface air temperature (°C) averaged over the Eurasian continent (from Yasunari et al., 1989, Fig. 1).

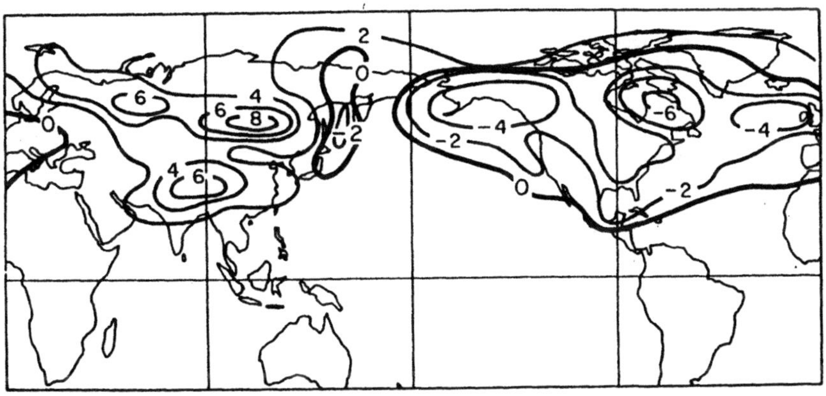

Figure 5. Differences of June sea level pressure (mb), four-case average of "doubled-snowfall" simulations minus the control simulation (from Barnett et al., 1989, Fig. 9).

Figure 6. Differences in monthly precipitation (cm) between the four-case average of the snow anomaly runs (H = Halved snowfall rate, D = Doubled snowfall rate) and the control run, C (from Barnett et al., 1989, Fig. 11).

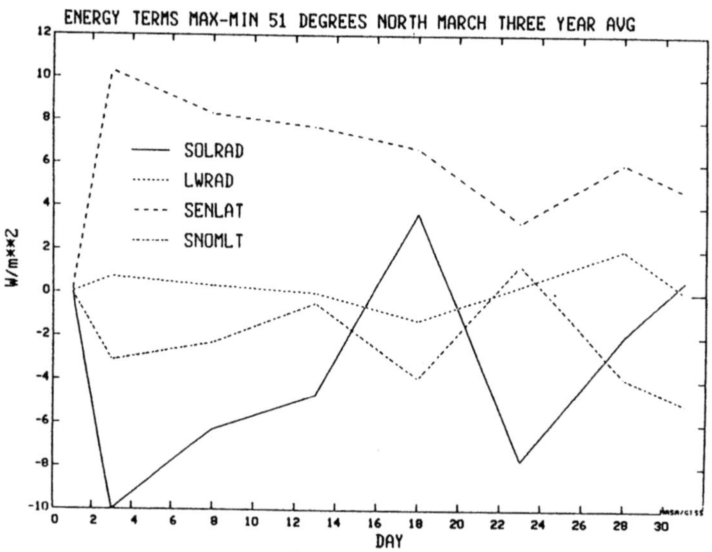

Figure 7. Latitudinally and five-day averaged differences ("maximum snow" minus "minimum snow") of the following terms of the surface energy budget: absorbed solar radiation (solid line), emitted longwave radiation (dotted line), combined sensible and latent heat (dashed line), and snow melt converted into equivalent energy units (uneven dashed line). Averages are for March 1978, 1979, and 1980 at 51°N (from Cohen and Rind, 1991, Fig. 8).

address the snow-ENSO link in more detail, Barnett et al. performed additional experiments with a coupled ocean-atmosphere model. The conclusion was that "the snow/monsoon signal has all the characteristics necessary to trigger the Pacific portion of an ENSO event, but the signal is too small by a factor of at least two . . .. In balance, it appears that snow-induced monsoon perturbations may be one of the (multiple) triggers that can initiate an ENSO cycle" (Barnett et al., 1989, p. 683-684). The elimination of major model biases and the improvement of the model resolution are clearly high priorities in the context of this large-scale signal involving snow cover.

The studies summarized above suggest that snow cover may play a major climatic role regionally and perhaps even globally. However, the most recently published investigation of snow-climate interactions provides some indications of a negative feedback involving snow anomalies, at least during late winter and early spring. Cohen and Rind (1991) used the GISS 9-layer model with $8^\circ \times 10^\circ$ resolution to examine the sensitivities to snow cover during March. The initial (1 March) boundary conditions corresponded to the observationally-derived "maximum" and "minimum" snow cover and depth over North America and Eurasia. A key feature of this experiment was that the snow cover was interactive rather than prescribed. In the results of a five-case sample of March simulations, the positive snow anomalies caused only a short-term local decrease in surface temperature. There was no non-local signal, and even the local signal became weak after ~7 days. In the model physics, the reduction in the surface absorption of solar radiation and the increased consumption of latent heat for melting the snow contributed to lower temperatures. However, the remaining terms in the surface energy budget (e.g., longwave radiation, sensible heat flux) adjusted so that they offset the cooling (Fig. 7). Thus a negative feedback limited the impact of the snow anomalies to a slight cooling of ~1°C--a smaller effect than indicated by the observational studies and by the model experiment of Yasunari et al. (1989). This negative feedback that limits the albedo effect of snow was also found in Barnett et al.'s (1989) first experiment, and the weak signal is not inconsistent with the results of Robock and Tauss (1986). Thus the relatively negligible impact of snow cover in the Cohen and Rind experiment may be due to the fact that their simulations did not extend into the spring season, when the hydrologic role snow anomalies can become more important. The compatibility of the results from these various model experiments nevertheless requires further attention.

3.3 Decadal timescales

Because the focus of this discussion is the prediction of interannual fluctuations, the role of snow cover in a longer-term climatic context will be noted only briefly. Specifically, we will mention several investigations of longer-term changes that bear upon the interannual applications surveyed above. Some relevant background with regard to climate modeling is provided by the Intergovernmental Panel on Climate Change, which notes that several models (of which most have also been used in interannual applications) achieve a "broadly realistic simulation of snow cover . . .. However, there are significant errors in the snow cover on regional scales in all models" (IPCC, 1990, p. 113). The model climatologies of snow cover should therefore be examined critically in any investigations of model sensitivity to snow cover.

The most critical parameter in model experiments involving snow cover is likely to be the surface albedo, which is parameterized rather crudely in many of today's GCMs. In reality, the albedo of snow-covered land is a complicated function of snow depth and age, terrain, vegetative cover, and even cloudiness and solar zenith angle. The inclusion of a simple temperature-dependence in the parameterization of snow albedo can change substantially the globally-averaged surface temperature increase due to a doubling of CO_2 (Washington and Meehl, 1986). Williams (1975) found that changes in surface albedo (and SST) influenced the ice-age atmospheric circulation of the NCAR model more than did the orographic changes caused by glaciation. In view of the somewhat disparate model sensitivities to snow albedo reported by Yasunari et al. (1989), Barnett et al. (1989) and Cohen and Rind (1991), the parameterization of surface albedo and model sensitivities to this parameterization appear to merit more attention in global modeling.

Finally, there are indications that the continental snow cover of recent years is less extensive than 10-20 years ago (Robinson and Dewey, 1990). Perhaps coincidentally, the increase of surface temperature over the last few decades is larger over land than over the oceans. If these changes persist or become larger, there are several implications for the forecasting of interannual variations. First, the "normals" used to define the snow anomalies will need to be modified accordingly. Second, investigations of sensitivities to interannual variations of snow cover may assume greater relevance to climatic change, especially if the more valid results of the studies surveyed in Section 3.2 are those indicating the stronger sensitivities to snow cover.

4. Conclusion

Several conclusions emerge from the results surveyed here. The first, which pertains to the strategy for diagnosing the role of snow cover in short-term climate variability, is that both modeling and data analysis are essential and complementary diagnostic tools. Because snow is involved in a variety of interactions within the climate system, and because the distribution of snow is determined largely by other components of the climate system, controlled experimentation with numerical models is a key element of the diagnostic strategy. However, the biases and other limitations of models are such that numerical experiments yield convincing conclusions only when the observational data give some credibility to the model results.

The diagnostic studies of the past few decades appear, at first glance, to have produced a somewhat inconclusive picture of the short-term climatic role of snow cover. However, the aggregate of the results does point to several scientific conclusions:

(a) Anomalies of snow cover are clearly associated with significant local anomalies of air temperature, at least in the lowest 100-200 mb. The duration of these local anomalies generally ranges from several days to several weeks, and is often limited by the fluctuations of the snow anomalies themselves over weekly and monthly timescales.

(b) There is little or no evidence that the albedo effect of wintertime snow anomalies produces meaningful signals in the large-scale atmospheric circulation.

(c) The snow-hydrology-soil moisture feedback does appear to be capable of producing a meaningful response in the atmosphere during spring and summer. The scale of this response is at least regional and possibly larger. Regions susceptible to snow-hydrology effects include eastern Asia and the northern portions of Eurasia and North America.

(d) The mechanism(s) by which snow influences the large-scale atmosphere are sufficiently complex that relatively sophisticated (e.g., nonlinear) models are required for the diagnosis of snow-atmosphere interactions.

With regard to (c) and (d), the validity of parameterizations involving snow cover in global climate models may be a major constraint on further progress toward an understanding of snow-atmosphere feedbacks. Priority should be given to the parameterization of quantities and processes such as snow surface albedo, the influence of snow on the boundary layer structure, and the disposition of the liquid water produced by snow melt.

Acknowledgements

The preparation of this paper was supported by NASA's Interdisciplinary Research Program through Grant IDP-88-009 and by the National Science Foundation through Grant ATM-8912911. We thank Norene McGhiey for typing the manuscript.

References:

Afanas'eva VB, Esakova NP, Klimentova RV. (1979) Relation of the planetary upper-air frontal zone to the position of the snow limit during fall and spring. Meteorologiya i Gidrologiya 9(1979):110-112

Barnett TP, Dümenil L, Schlese U, Roeckner E, Latif M (1989) The effect of Eurasian snow cover on regional and global climate variations. J Atmos Sci 46:661-685

Blanford HF (1884) On the connection of the Himalayan snowfall with dry winds and seasons of drought in India. Proc Royal Soc London 37:3-32

Bhanu Kumar OSRU (1988) Interaction between Eurasian winter snow cover and location of the ridge at the 500 hPa level along 75°E. J Meteorol Soc Japan 66:509-514

Bhanu Kumar OSRU (1987) Seasonal variation of Eurasian snow cover and its impact on the Indian summer monsoon. Large Scale Effects of Seasonal Snow Cover (Goodison BE, Barry RG, Dozier J, Eds), IAHS Press, Wallingford UK, 51-60

Cohen J, Rind D (1991) The effect of snow cover on the climate. J Climate 4: in press

Dewey KF (1977) Daily maximum and minimum temperature forecasts and the influence of snow cover. Mon Weather Rev 105:1594-1597

Dey B, Bhanu Kumar OSRU, Kathuria SN (1984) Himalayan snow cover and the Indian summer monsoon activity. Final Report, NSF Grant ATM-8109177, Howard Univ, Washington DC, 30 pp

Dickson RR (1984) Eurasian snow cover versus Indian monsoon rainfall--An extension of the Hahn-Shukla results. J Clim Appl Meteor 23:171-173

Dickson RR, Namias J (1976) North American influence on the circulation and climate of the North Atlantic sector. Mon Weather Rev 104:1255-1265

Foster J, Owe M, Rango A (1983) Snow cover and temperature relationships in North America and Eurasia. J Appl Meteor 22:460-469

Hahn DG, Shukla J (1976) An apparent relationship between snow cover and Indian monsoon rainfall. J Atmos Sci 33:2461-2462

IPCC (1990) Climate Change: The IPCC Scientific Assessment. Intergovernmental Panel on Climate Change, WMO/UNEP, Cambridge Univ Press, Cambridge UK, 365 pp

Iwasaki T (1991) Year-to-year variation of snow cover area in the Northern Hemisphere. J Meteorol Soc Japan 69:209-217

Lamb HH (1972) Climate: Past, Present and Future, Vol. 1. Fundamentals and Climate Now. Methuen, London, 613 pp

Lamb HH (1955) Two-way relationship between snow or ice limit and 1000-500 mb thickness in the overlying atmosphere. Quart J Royal Meteorol Soc 81:172-189

Namias J (1985) Some empirical evidence for the influence of snow cover on temperature and precipitation. Mon Weather Rev 113:1542-1553.

Namias J (1981) Snow covers in climate and long-range forecasting. Glaciological Data, World Data Center A for Glaciology, Boulder CO, GD-11:13-26

Namias J (1962) Influence of abnormal heat sources and sinks on atmospheric behavior. Proceedings, Symposium on Numerical Weather Prediction, Meteorol Soc of Japan, Tokyo, 615-627

Opsteegh JD, Mureau R (1984) Description of a 15-layer steady-state atmospheric model. Report SF-84-19, Dept of Meteorol, Univ Maryland, 25 pp

Pan HL, Campana KA, Kanamitsu M (1989) Initial data for snow, sea ice, soil temperature, soil moisture, and convective cloud in NMC's global model. Research Activities in Atmospheric and Oceanic Modelling (G Boer, Ed) WMO/TD-No 332, 6.14-6.15

Petersen RA, Hoke JE (1989) The effect of snow cover on the Regional Analysis and Forecast System (RAFS) low-level forecasts. Weather and Forecasting 4:253-257

Roads JO (1981) Linear and nonlinear aspects of snow albedo feedbacks in atmospheric models. J Geophys Res 86:7411-7424

Robinson DA, Dewey KF (1990) Recent secular variations in the extent of Northern Hemispheric snow cover. Geophys Res Lett 17:1557-1560

Robock A, Tauss JW (1986) Effects of snow cover and tropical forcing on mid-latitude monthly mean circulation. Glaciological Data, World Data Center A for Glaciology, Boulder, CO, GD-18:207-214.

Ropelewski CF, Robock A, Matson M (1984) Comments on "An apparent relationship between Eurasian spring snow cover and the advance period of the Indian summer monsoon. J Clim Appl Meteorol 23:341-342

Ross B, Walsh JE (1986) Synoptic-scale influences of snow cover and sea ice. Mon Weather Rev 114:1795-1810

Toomig KG (1981) Correlation of mean annual albedo and short-wave radiation balance with these parameters in early spring. Meteorologiya i Gidrologiya 5(1981):48-52

Treidl RA (1970) A case study of warm air advection over a melting snow surface. Boundary Layer Meteorol 1:155-168

Wahl EW (1968) A comparison of the climate of the eastern United States during the 1830's with the current normals. Mon Weather Rev 96:73-82

Walker GR (1910) Correlations in seasonal variations of weather. Mem India Meteorol Dept 21:22-45

Walsh JE, Jasperson WH, Ross B (1985) Influences of snow cover and soil moisture on monthly air temperature. Mon Weather Rev 113:756-768

Walsh JE, Ross B (1988) Sensitivity of 30-day dynamical forecasts to continental snow cover. J Climate 1:739-754

Washington WM, Meehl GA (1986) General circulation model CO_2 sensitivity experiments: Snow-sea ice albedo parameterizations and globally averaged surface air temperature. Clim Change 8:231-241

Wiesnet DR, Matson M (1976) A possible forecasting technique for winter snow cover in the Northern Hemisphere and Eurasia. Mon Weather Rev 104:828-835

Williams J (1975) The influence of snowcover on the atmospheric circulation and its role in climatic change: An analysis based on results from the NCAR global circulation model. J Appl Meteorol 14:137-152

Yasunari T, Kitoh A, Tokioka T (1989) The effect of Eurasian snow cover on the summer climate of the Northern Hemisphere: A study by the MRI-GCM. Research Activities in Atmospheric and Oceanic Modelling (G Boer, Ed) WMO/TD-No 332, 7.37-7.38

Yeh TC, Wetherald R, Manabe S (1983) A model study of the short term climatic and hydrological effects of sudden snow-cover removal. Mon Weather Rev 111:1013-1024

COUPLED TOGA MODELS AT THE UKMO

M. Davey*, S. Lawrence
Robert Hooke Institute
Clarendon Laboratory
Parks Road Oxford OX1 3PU
United Kingdom

C. Gordon, S. Ineson
Hadley Centre
UK Meteorological Office
London Road
Bracknell Berks. RG12 2SY
United Kingdom

* Meteorological Office Unit

Abstract

Both general circulation (GCM) and intermediate ocean and atmosphere models are being used by the tropical ocean modelling group at the UK Meteorological Office to investigate interannual variability. A 13-year coupled GCM integration was completed in 1990, producing a realistic seasonal cycle but little interannual variability. Some details of year 8 of that integration are presented here, and a related hindcast experiment is discussed. An intermediate model with two active layers in the ocean and an equilibrium atmosphere is described: ocean climatology and preliminary coupled experiments are discussed.

1. Introduction

As part of the Tropical Ocean Global Atmosphere (TOGA) programme several numerical models are being developed and used jointly by the tropical ocean groups at the Hadley Centre at the United Kingdom Meteorological Office (UKMO) in Bracknell, and at the Hooke Institute (HI) in Oxford. The work is focussed on the Pacific region, and on the interannual variability associated with El Niño Southern Oscillation (ENSO)

NATO ASI Series, Vol. I 6
Prediction of Interannual Climate Variations
Edited by J. Shukla

events, with the aim of both understanding the mechanisms of such events and of predicting their occurrence.

ENSO events involve strong interaction of the ocean and atmosphere. It is expected that the most realistic numerical simulations of past events, and predictions of future events, will eventually be based on the detailed general circulation models that at present best represent the oceanic and atmospheric components when considered in isolation. Coupling these components together is far from straightforward however: taken separately one component behaves well when tied to the observed properties of the other, but a coupled model can easily drift into an unreasonable state when released from the observational ties. Certainly that was the case for the first coupled experiments at UKMO (Gordon 1989) and elsewhere, and subsequent work has concentrated on eliminating or neutralising such drift. For the UKMO coupled model the main early error was overheating of the western Pacific ocean that led to a permanent state like an El Niño. The specific addition of gustiness associated with convection and other general improvements to the atmospheric GCM largely overcame that defect, and led to the improved simulation described in section 2 below. A basic intercomparison of this and other coupled models has been carried out by Neelin et al. (1991).

The coupled GCMs have not yet reached the stage of making predictions, and their expense and complexity make them awkward tools to use for process studies. This gap is filled by intermediate models, which retain the main dynamical processes thought to be relevant to ENSO but in a greatly reduced form. These models reveal various ways that ENSO events may begin and evolve (see e.g. Anderson & McCreary (1985), the linear model by Hirst (1986), Schopf & Suarez (1988) and others), and when used to describe perturbations from climatology such models can have substantial predictive skill (notably Cane et al. (1986)). The first such UKMO/HI model had a single-active-layer ocean and a time-dependent atmosphere, and produced eastward propagating coupled interannual modes. In a version with active Pacific, Atlantic and Indian oceans this interannual activity was found to be confined to the Pacific region, and was strongly influenced by conditions over the land masses (Budin & Davey 1990). As a continuation of that project, a two-layer ocean with better climatology has been developed as described in section 3 below, and initial experiments with a simple equilibrium atmosphere have been carried out.

2. The coupled GCM

The present TOGA coupled GCM used at the UKMO has been developed over a number of years. The atmospheric component is the standard global one used for climate modelling at the UKMO, with 2.5° x 3.75° latitude-longitude resolution, and

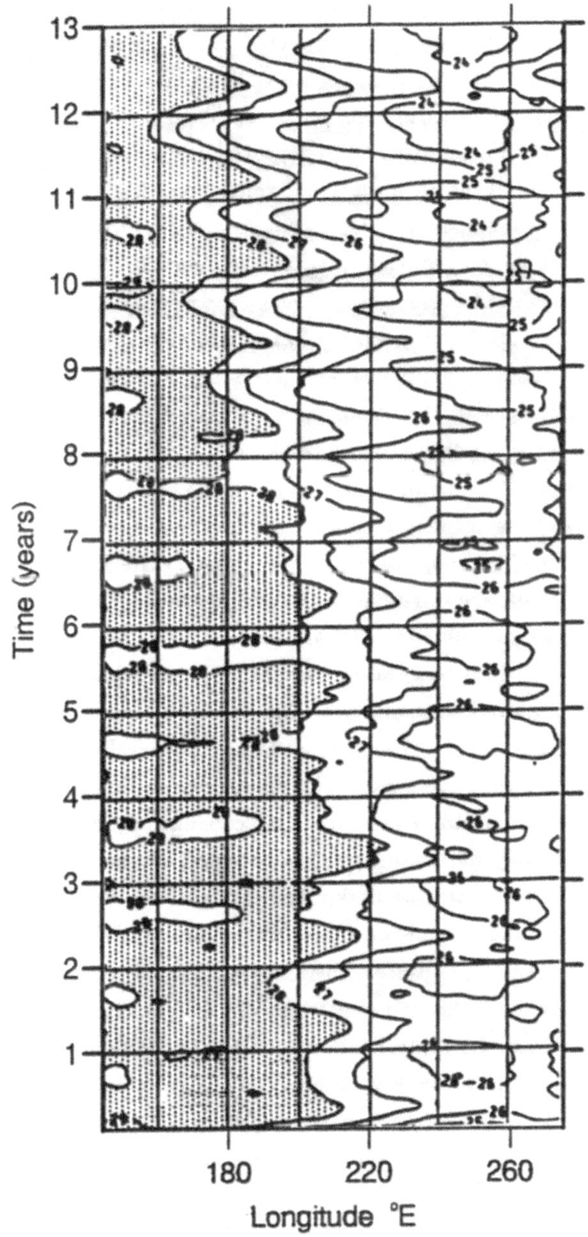

Fig. 1: Pacific equatorial sea surface temperature (°C), averaged over 5N to 5S, for 13 years from the coupled GCM. The seasonal cycle is realistic, but there is a gradual cooling trend. Temperatures greater than 28°C are stippled.

11 vertical levels. The version used for the experiment described here includes effects such as interactive cloud specified via an explicit cloud liquid water scheme: see Slingo (1985) for a description of the basic model, and Slingo et al. (1989) and Smith (1990) for recent improvements. The ocean component is a version of the Bryan & Cox model (Cox 1984), covering the Pacific from 30°N to 30°S. Meridional resolution varies from 0.33° at the equator to 1° at the open north and south boundaries, while zonal resolution is 1.5°, reducing to 0.5° near the coasts, and there are 16 vertical levels. Simulations using the ocean model alone have been described by Gordon & Corry (1991).

The coupled experiment presented here was carried out by Gordon and Ineson in 1990. For the ocean component an energy balance mixed-layer model was embedded (c.f. Kraus & Turner 1967), to supplement the K-theory scheme (Pacanowski & Philander 1981) used previously. For the atmosphere, evaporation in convective regions was increased by adding a parameterisation of gustiness (Gordon, private communication).

The ocean is first spun up from rest, from initial climatological temperature and salinity (Levitus 1982), using wind stresses and heat fluxes obtained from the atmospheric GCM forced by climatological sea surface temperature (SST). Surface temperature and salinity are kept close to climatological values during the spinup period of over 3 years. The ocean is then fully coupled to the atmosphere. No 'flux correction' terms to overcome climate drift (see e.g. Sausen et al. 1988) are included.

Fig. 1 shows equatorial sea surface temperature T (averaged from 5°N to 5°S) during the coupled integration. The seasonal cycle is generally quite well simulated, with a pattern close to that observed. Temperature variation in the east Pacific is somewhat weak however, and there is a long-term cooling trend indicating slow climate drift. The ocean is initially too warm, and has become too cool by the end of the run. There is also some inter-annual variability, but much smaller than that associated with observed ENSO events, and there is no sign of spontaneous ENSO development.

Although no ENSO event appeared, this integration does represent a self-consistent ocean-atmosphere simulation that behaved well. SST for year 8 is close to climatological values, so we shall consider here some aspects of the simulation for that year.

From the difference shown in Fig. 2 between the equatorial model SST and the observed monthly SST climatology (here denoted T_c, taken from the COADS dataset by Woodruff et al. 1987) we see that the eastern cold tongue in the model is too cool in northern spring, and too warm in autumn. The observed SST is warmer in spring than in autumn, so this difference reflects a lack of model variability in this region, which is related to weakness of the model SE trade winds there, particularly in autumn. Further, there is little variation in the depth of the 20°C isotherm, which fluctuates between $60m$ and $80m$ in the equatorial east Pacific.

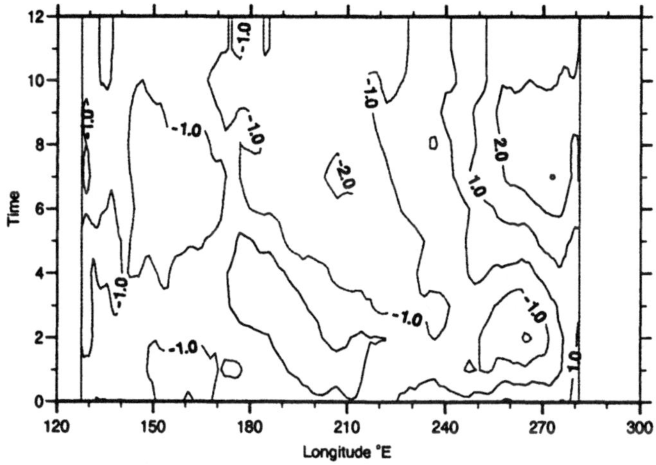

Fig. 2: Equatorial sea surface temperature difference: coupled GCM (year 8) minus climatological values from the Comprehensive Ocean-Atmosphere Data Set. Minimum −2.2°C, maximum 3.0, contour interval 1.

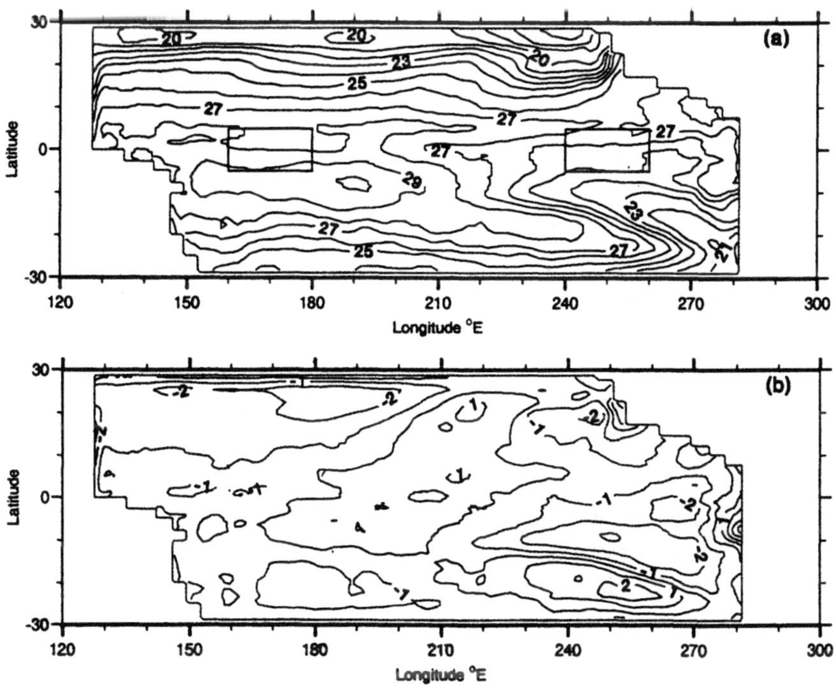

Fig. 3: Ocean surface temperatures for March of year 8 from the coupled GCM. (a) model SST, maximum 30.2°C, minimum 16.4, contour interval 1. The rectangular regions drawn in the west and east Pacific indicate areas used for the scatter plots in Figs. 4 and 5. (b) SST difference (model - COADS climatology), maximum 4.5 on South American coast, minimum -3.4, contour interval 1.

The spatial patterns of T and $T - T_c$ are given for March in Fig. 3. The difference is small over most of the Pacific region, with the largest (warm) anomaly appearing off the coast of S. America, and errors of $\pm 2°$ in the SE Pacific.

The net heat flux Q into the ocean is generally small in the west Pacific (typically $20Wm^{-2}$), increasing to over $100Wm^{-2}$ in parts of the equatorial east Pacific. Fig. 4 gives scatter plots of Q vs. T for the two equatorial regions outlined in Fig. 3a, one just west of the dateline and the other in the east Pacific. For the western region (Fig. 4a) the relation is surprisingly linear, with a dQ/dT slope of about $-35Wm^{-2}K^{-1}$, whereas in the eastern region (Fig. 4b) with cooler SST there is greater scatter with a tendency for positive (negative) dQ/dT for lower (higher) temperatures. Scatter plots for the heat flux components were also made: Fig. 5 shows the sensible+latent+infrared flux (dominated by the latent heat component), and the net solar flux vs. SST, for the western region. Where SST is above 28°C cloud shading (not shown) reduces the solar component substantially, but this is compensated by reduced evaporative cooling, despite the increased gustiness factor associated with precipitation there. The near-linear net relation may be fortuitous: there is somewhat more scatter in the corresponding September plot (not shown) for example. K.-M. Lau (private communication) has suggested that the relation could also be a property of the type of convection scheme used in the atmospheric GCM.

Using the above integration as a control, hindcasts of the 1982-83 ENSO event using the coupled GCM were attempted. In previous experiments the Pacific ocean GCM alone had been spun up using climatological winds (taken from the atmospheric GCM alone with climatological SST), and also using windstress for the 1980s provided by Florida State University (called FSU pseudostress, c.f. Goldenberg & O'Brien 1981). For the hindcasts the ocean component was initialised by calculating the September 1982 difference between the ocean-only integrations with climatological and FSU stresses, and adding that difference to the ocean state in the coupled integration at September of year 8. In this way a realistic complete ocean anomaly is added to a balanced ocean-atmosphere system, giving a substantial warm SST anomaly in the central Pacific. In the 82/83 ENSO event this anomaly was observed to spread eastward in the following months. In the model hindcast experiment however the warm anomaly rapidly decayed, due to strong easterly surface winds induced east of that region that caused strong surface cooling. The general structure of the atmospheric response to the anomaly was unsatisfactory, with quite different wind anomalies appearing above and within the boundary layer, for reasons that remain unclear. For some hindcasts the atmosphere was allowed to adjust to the new surface conditions before coupling, but results were essentially the same as were found without this period of adjustment.

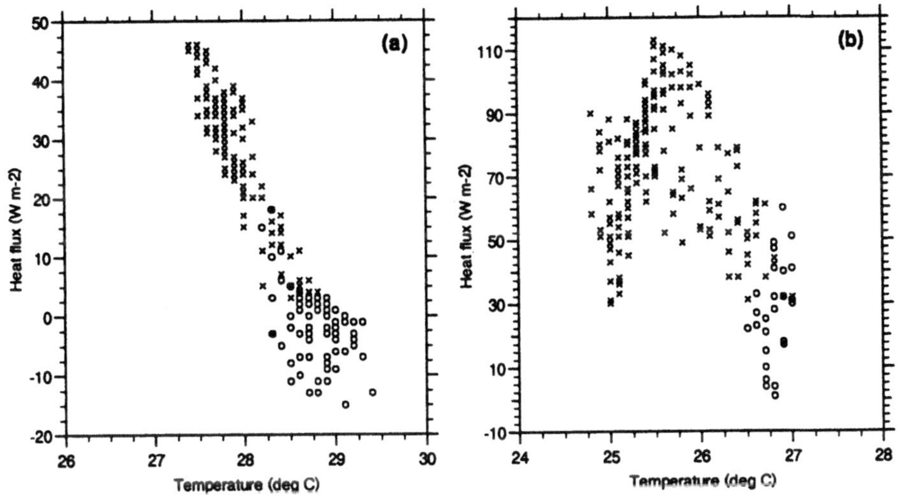

Fig. 4: Scatter plots of net surface heat flux Q into the ocean against sea surface temperature, for the March average of year 8 from the coupled GCM, for the regions shown in Fig. 3a. The symbol o is used where the model rainfall rate is greater than 10 mm/day, × otherwise, for ocean model gridpoints in the region. (a) In the west Pacific area (5S to 5N, 160E to 180E) the relation is almost linear, with a slope of $-35Wm^{-2}K^{-1}$. (b) For the east Pacific area (5S to 5N, 240E to 260E) there is more scatter, with Q increasing with SST at the lower end of the SST range.

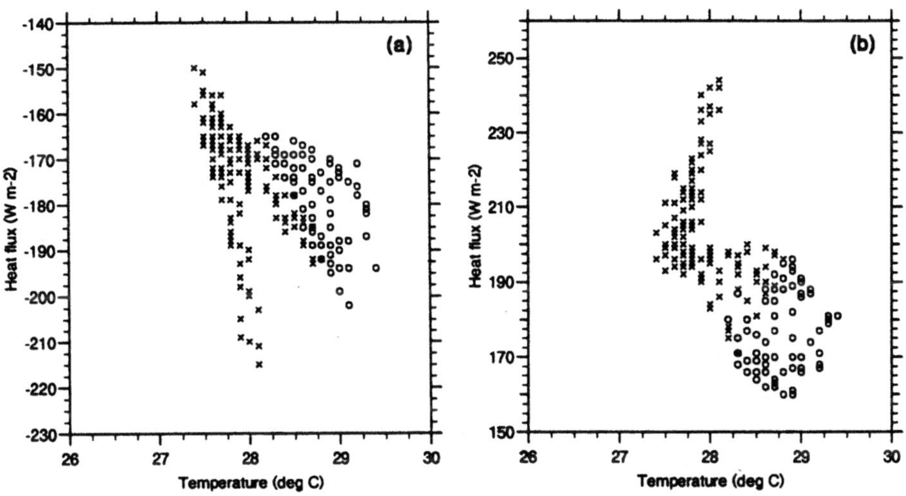

Fig. 5: Scatter plots of components of the heat flux Q into the ocean against sea surface temperature, for the March average of year 8 from the coupled GCM, for the west Pacific region 5S to 5N, 160E to 180E. The symbol o is used where the model rainfall rate is greater than 10 mm/day, × otherwise, for ocean model gridpoints in the region. (a) Sensible+latent+infrared vs. SST: the latent heat flux term is the main component. (b) Solar vs. SST: the solar component decreases where SST is largest because precipitation and cloud shading is active there.

3. Intermediate model

The reduced-physics intermediate coupled model of Budin & Davey (1991) contained an ocean with a thick single active layer. The thickness of the layer made the thermal response of the ocean rather sluggish, as found in experiments with added high frequency noise in the atmosphere, and model ocean currents could only represent an average over the depth of that layer. A new ocean model with two active layers, which could represent both surface currents and undercurrents, is being developed as described below.

3.1 Ocean equations

The ocean has two active layers with varying depth, linearised about a basic state with layer depths $\bar{h}_1$ and $\bar{h}_2$, and temperatures $\bar{T}_1$ and $\bar{T}_2$. Below the second layer the temperature is $\bar{T}_3$. The ocean dynamic equations are standard (c.f. the nonlinear model by McCreary & Kundu 1988):

$$\frac{\partial \underline{u}_1}{\partial t} + f\underline{k}\mathrm{x}\underline{u}_1 = -\frac{1}{\rho_0}\nabla p_1 + \nu\nabla^2\underline{u}_1 - d_1(\underline{u}_1 - \underline{u}_2) + \frac{\underline{\tau}}{\rho_0 \bar{h}_1} \quad (1)$$

$$\frac{\partial \underline{u}_2}{\partial t} + f\underline{k}\mathrm{x}\underline{u}_2 = -\frac{1}{\rho_0}\nabla p_2 + \nu\nabla^2\underline{u}_2 + d_2(\underline{u}_1 - \underline{u}_2) \quad (2)$$

$$\frac{\partial h_1}{\partial t} + \bar{h}_1\nabla.\underline{u}_1 = w_{e1} + w_{d1} \quad (3)$$

$$\frac{\partial h_2}{\partial t} + \bar{h}_2\nabla.\underline{u}_2 = -w_{e1} - w_{d1} + w_{e2} + w_{d2} \quad (4)$$

An equatorial beta-plane $f = \beta y$ is used. The positive terms w_{e1} (entrainment from layer 2 into layer 1) and w_{e2} (from the deep ocean into layer 2) are used to limit shallowing of each layer. They act only when a layer depth is less than some critical value h_e, giving entrainment that is roughly proportional to the square of $h_e - h$. Entrainment into the upper layer may be regarded as a crude parameterisation of the effect of wind mixing penetrating into layer 2. Similarly, negative detrainment terms w_{d1} and w_{d2} are active when layer depths exceed critical depths h_d, and are needed to allow the total volume of a layer to reach equilibrium by returning water gained by entrainment. The entrainment/detrainment algorithm also allows detrainment from layer 1 to layer 2 when layer 2 is too shallow, in preference to entrainment from the deep ocean.

Temperature is held fixed in the pressure gradient terms:

$$\frac{1}{\rho_0}\nabla p_1 = g\alpha[(\bar{T}_1 - \bar{T}_2)\nabla h_1 + (\bar{T}_2 - \bar{T}_3)\nabla(h_1 + h_2)] \tag{5}$$

$$\frac{1}{\rho_0}\nabla p_2 = g\alpha(\bar{T}_2 - \bar{T}_3)\nabla(h_1 + h_2) \quad . \tag{6}$$

Thus $\underline{u}_1$, $\underline{u}_2$, h_1 and h_2 are determined entirely by the windstress, and are independent of any temperature variations. The windstress is calculated from either monthly climatology (0.75 Hellerman stress, Hellerman & Rosenstein 1983), or FSU pseudostress multiplied by drag coefficient C_D and air density $1.2 Kg\ m^{-3}$.

The flux-form thermodynamic equation for layer 1 is

$$\frac{\partial(h_1 T_1)}{\partial t} + \nabla.(\bar{h}_1 T_1 \underline{u}_1) = \frac{Q}{\rho_0 c_p} + A_T h_1 \nabla^2 T_1 + w_{e1} T_2 + w_{d1} T_1 \quad . \tag{7}$$

Originally the thermodynamic equation for layer 2 had the same form:

$$\frac{\partial(h_2 T_2)}{\partial t} + \nabla.(\bar{h}_2 T_2 \underline{u}_2) = A_T h_2 \nabla^2 T_2 - w_{e1} T_2 - w_{d1} T_1 + w_{e2} T_3 + w_{d2} T_2 \quad . \tag{8}$$

This form was chosen to conserve the total heat content of the two layers in the absence of Q, w_{e2} and w_{d2}, and to allow undercurrent advection of warm water from the west Pacific to influence the temperature of water upwelled in the east Pacific. The model took many years to reach thermal equilibrium however, so the layer 2 temperature was instead held fixed in the cases described below, with $T_2 = 23°C$. Effectively this fixes the temperature of water entrained into layer 1; T_2 has no other role in this model.

The equation governing T_1 can also be expressed as

$$\frac{\partial(T_1)}{\partial t} + (\bar{h}_1/h_1)\underline{u}_1.\nabla T_1 = \frac{Q}{\rho_0 c_p h_1} + A_T \nabla^2 T_1 + w_{e1}(T_2 - T_1)/h_1 \quad , \tag{9}$$

and this is the version is used in numerical integrations. The surface heat flux Q is linearised about climatological values Q_c, with

$$Q = Q_c - \left(\frac{dQ}{dT}\right)_c (T_c - T_1) \quad , \tag{10}$$

where Q_c and sea surface temperature T_c are taken from the COADS monthly climatology and $(dQ/dT)_c$ is as calculated by Oberhuber (1988).

The equations are solved on a variable C-grid for the tropical Pacific ocean, with uniform $\Delta x = 1.5°$ longitude and Δy varying from $\Delta y_0 = 0.59°$ at the equator to $2°$ at the solid walls near 30°N and 30°S. (The latitudinal resolution is just half that used for the ocean GCM in the coupled experiment described in section 2.) The viscosity varies as $A_M = A_{M0}\Delta y/\Delta y_0$.

Parameters for the experiments described below are:

$\bar{T}_1 = 28°C$	$\bar{T}_2 = 20°C$	$\bar{T}_3 = 12°C$
$\bar{h}_1 = 100m$	$\bar{h}_2 = 150m$	
$h_{1e} = 40\ m$	$h_{2e} = 40\ m$	
$h_{1d} = 200\ m$	$h_{2d} = 500\ m$	
$d_1 = 5.8\mathrm{x}10^{-8} sec^{-1}$	$d_2 = (\bar{h}_1/\bar{h}_2)d_1$	
$A_M = 2000m^2 sec^{-1}$	$A_T = 2000m^2 sec^{-1}$	
$C_D = 1.15\mathrm{x}10^{-3}$	$\alpha = 2.5\mathrm{x}10^{-4}$	

The characteristic wave speeds for the two internal modes are $c_1 = 2.43msec^{-1}$ and $c_2 = 0.99msec^{-1}$. Generally only mass exchange between layer 1 and layer 2 is active: under these conditions the faster mode is very lightly damped, while the slower mode is strongly damped in regions of active upwelling.

3.2 Ocean climatology

The model was forced for several years, using 0.75 of the monthly Hellerman climatology for the wind stress $\underline{\tau}$. Fig. 6 shows the resulting equatorial temperature cycle for one year, for T_1 and $T_1 - T_c$. In the central and west Pacific there is no upwelling into layer 1. Advection effects are small in the western region, and there T_1 is effectively driven locally by the surface heat flux. In the east Pacific, where the upper layer is shallow and entrainment w_{e1} is active, upwelling of 23° water counteracts the surface heat flux to maintain the equatorial cold tongue. This cool water is then advected westward into the central region.

The simple thermodynamics used here, with the temperature of upwelled water kept constant, tends to produce eastern SST that is up to 2° too cool in northern spring (Fig. 6b) when upwelling is strongest. A warm patch appears around 230E (130W) in May and June, but then rapidly fades to be replaced by a weak cool anomaly.

The spatial SST structure at the end of March is shown in Fig. 7: temperatures in the equatorial region at this time are generally close to T_c. The largest tropical anomalies are in the eastern region, with the narrow equatorial cold tongue being about 1° too cool, and substantial warm anomalies next to the S. American coast. The corresponding currents $\underline{u}_1$ and $\underline{u}_2$ are shown in Fig. 8, with prominent zonal equatorial currents and undercurrents with magnitude about $0.3m\ sec^{-1}$, similar to the observed average currents (Wyrtki & Kilonsky 1984).

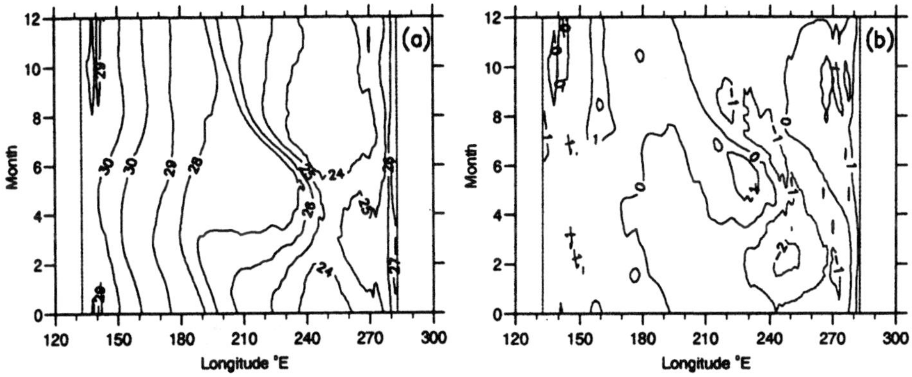

Fig. 6: The annual cycle in equatorial sea surface temperature for the intermediate ocean model forced by 0.75 Hellerman wind stress. (a) T_1, maximum 30.3°C, minimum 23.2, contour interval 1. (b) The difference $T_1 - T_c$, where T_c is calculated from the COADS climatology. Minimum -2.3, maximum 1.8, contour interval 0.5 .

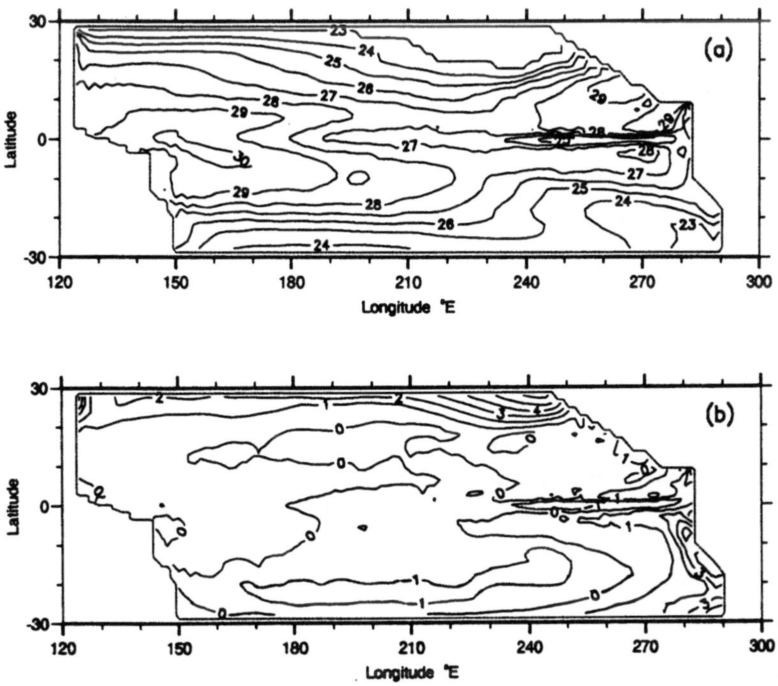

Fig. 7: Sea surface temperatures for April 1 from the intermediate ocean model forced by 0.75 Hellerman wind stress. (a) T_1, maximum 30.2°C, minimum 23.0, contour interval 1 . (b) The difference $T_1 - T_c$, where T_c is calculated from the COADS climatology. Maximum 6.3 in the north-east corner, minimum -2.0, contour interval 1.

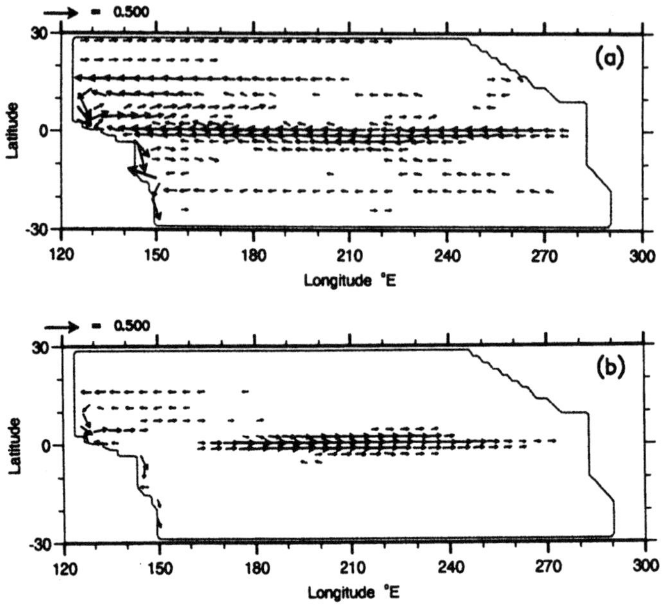

Fig. 8: Currents for April 1 from the intermediate ocean model forced by 0.75 Hellerman monthly wind stress climatology. (a) upper layer $\underline{u}_1$, maximum speed $0.45 m\ sec^{-1}$. (b) lower layer $\underline{u}_2$, maximum speed 0.42 .

3.3 The ocean response to 1980-1988 winds

The ocean model was next forced with the FSU winds, first by spinning up with an average of the 1980-1988 winds, then with the monthly winds from 1980 to 1988. The climatological heat flux formulation (10) was retained. The equatorial SST is given in Fig. 9: in 1982-1983 28° water spreads from the west across to the east, as westward surface currents weaken and the upper layer deepens in the east Pacific to restrict upwelling of cool water into the upper layer. The equatorial cold tongue reappears rapidly and strongly near the end of 1983. In 1987 warm water again extends to the east Pacific, but 28° water only appears briefly there. The general pattern is similar to the observed ENSO events, but eastern sea surface temperatures are not as high as observed.

The two ENSO events show more clearly in the evolution of upper layer depth perturbation along the equator (Fig. 10). In response to the weakening easterly winds, the upper layer rapidly deepens in the east to shut off entrainment there during the two events.

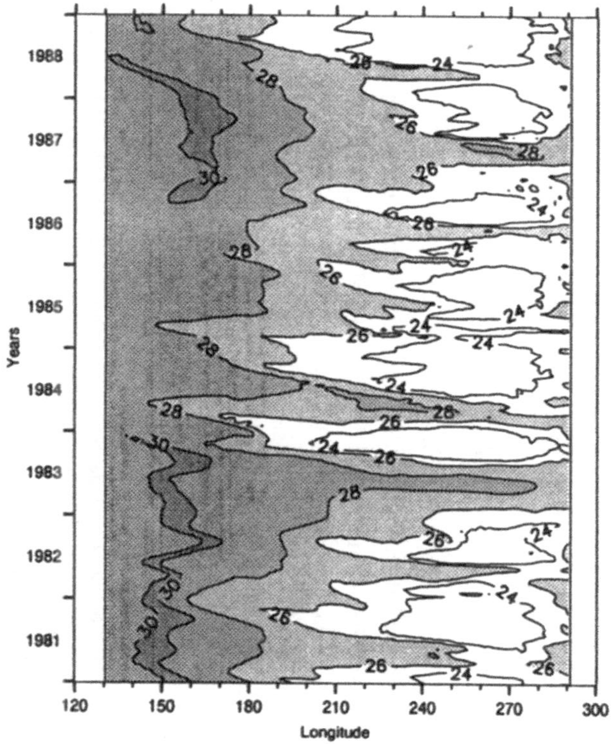

Fig. 9: Equatorial sea surface temperature T_1 for 1981 to 1988 from the intermediate ocean model forced by FSU pseudostress. Temperatures greater than 26°C are shown stippled. Maximum 30.3°C, minimum 23.0, contour interval 1.

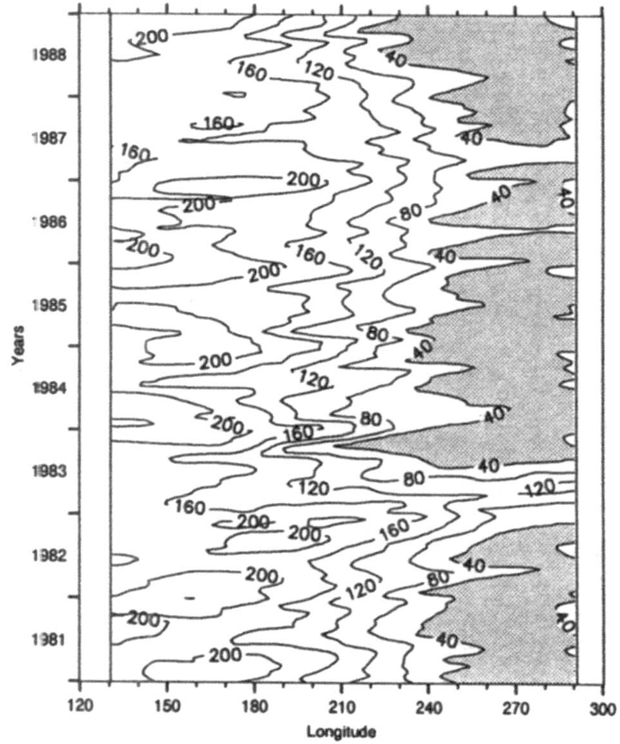

Fig. 10: Equatorial upper layer depth h_1 for 1981 to 1988 from the intermediate ocean model forced by FSU pseudostress. Maximum 240m, minimum 26m, contour interval 40m. Entrainment from the lower layer to this layer occurs only in the stippled area where $h_1 < 40m$.

3.4 Preliminary coupled experiments

Sea surface temperature perturbations from the seasonal cycle described in section 3.2 above were used to force a simple equilibrium atmosphere model, to obtain surface wind perturbations $\underline{U}'$. The equations for the atmosphere have the dry shallow-water form (c.f. Gill 1980):

$$f\underline{k}x\underline{U}' = -\nabla\Phi' - \epsilon\underline{U}' \qquad (11)$$

$$c^2\nabla.\underline{U}' = -Q_a{}' - \epsilon\Phi' \quad , \qquad (12)$$

with $c = 60msec^{-1}$ and $\epsilon = 4x10^{-6}sec^{-1}$.

In the case described below the perturbation atmospheric thermal forcing $Q_a{}'$ is determined only by sea surface temperature changes, with no convergence feedback. The forcing is

$$Q_a{}' = K_1[(T_1 - 28)_+ - (T_{cm} - 28)_+] + K_2(T_1 - T_{cm}) \quad , \qquad (13)$$

where T_{cm} is the model SST from the climatology experiment above, and the subscript + indicates that only positive values of the bracketed quantity are used. The term with coefficient K_1 represents heating by deep convection where SST exceeds $28°C$ (c.f. Anderson & McCreary 1985).

The wind stress perturbation thus obtained is added to the climatological stress used to force the ocean model:

$$\underline{\tau}' = K_T\underline{U}' \quad , \qquad (14)$$

with $K_T = 8.5x10^{-3}$, for stress in $N\ m^{-2}$ and wind in $m\ sec^{-1}$.

Experiments were carried out by coupling the perturbation atmosphere (every 10 days) after spinning up the ocean seasonal cycle. Residual spinup, and small errors in interpolating the saved values T_{cm}, gave small Q' for the first few months after coupling. Various values for K_1 and K_2 have been tried. For $K_1 = K_2 = 2x10^{-3}\ W\ m^{-2}K^{-1}$ perturbations remained negligibly small, whereas with $K_1 = K_2 = 8x10^{-3}$ perturbations grew to an unrealistic size, with strong coupled biennial oscillations. For $K_1 = K_2 = 4x10^{-3}$ small cold SST anomalies in the east caused easterly wind perturbations that in turn led to seasonal extension westward of the cold tongue, a little further each year. After 8 years of this behaviour the system settled into a regular annual cycle with substantial perturbations from the uncoupled climatology. Fig. 11 shows two cycles of equatorial SST in this case: large scale movement westward of cool water is prominent each year, with rapid changes evident across the whole Pacific. The system generally behaves like an amplified annual cycle biased toward a cold La Niña state.

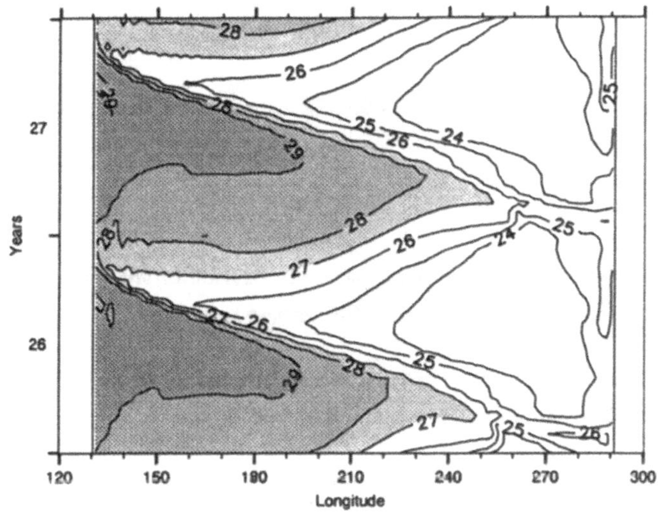

Fig. 11: Equatorial sea surface temperature T_1 for the intermediate ocean model coupled to a simple equilibrium atmosphere which is driven by perturbations from the SST climatology in Fig. 6. The coupled model has settled into an exaggerated annual cycle.

4. Future work

The present state of TOGA modelling work at the UKMO has been described. For the coupled GCM a new recoded atmospheric component will soon be available. Preliminary uncoupled SST anomaly (September 1982) tests with a trial version of the new atmosphere GCM show much better structure than was obtained in the hindcast attempts of section 2, with westerly surface wind anomalies occurring over the warm SST anomaly. The ocean GCM is being adapted for consistency with the new operating system, and ocean-only experiments will be carried out before new coupled GCM work.

Coupled experiments with intermediate models are continuing at the Hooke Institute. Allen & Davey (1991) have used an inverse atmosphere model to determine optimal parameterisations of the surface wind response to SST anomalies, and that scheme will be tested in a coupled model. Deng-Hua Wu (private communication) has added a third fixed-depth layer to a similar intermediate ocean, and if trials are successful then a third layer may be added to the ocean model described here. A hybrid model with a Pacific ocean GCM coupled to a simple equilibrium atmosphere has been developed by Stockdale (1991 D.Phil. thesis) to investigate coupled modes and to carry out ENSO hindcasts.

Acknowledgement

Ray McQuigg (UKMO) provided valuable assistance with data preparation.

References

Allen MR, Davey MK (1991) Empirical parameterisation of tropical ocean-atmosphere coupling: the 'inverse Gill' problem. Submitted to J Clim

Anderson DLT, McCreary JP (1985) Slowly propagating disturbances in a coupled ocean-atmosphere model. J Atmos Sci 42:615-628

Budin G, Davey MK (1990) Intermediate models of intraseasonal and interannual waves in the tropics. J Mar Sys 1:39-50

Cane MA, Zebiak SE, Dolan SC (1986) Experimental forecasts of El Niño. Nature 321:827-832

Cox MD (1984) A primitive equation, 3-dimensional model of the ocean. GFDL Ocean Group Tech Rep. 1, Geophys. Fluid Dyn. Lab./NOAA, Princeton, N.J.

Gill AE (1980) Some simple solutions for heat-induced tropical circulations. Q J Roy Met Soc 106:447-462

Goldenberg SB, O'Brien JJ (1981) Time and space variability of tropical Pacific wind stress. Mon Weath Rev 109:1190-1207

Gordon C (1989) Tropical ocean-atmosphere interactions in a coupled model. Phil Trans Roy Soc London A329:207-223

Gordon C, Corry RA (1991) A model simulation of the seasonal cycle in the tropical Pacific ocean using climatological and modeled surface forcing. J Geophys Res 96:847-864

Hellerman S, Rosenstein M (1983) Normal monthly windstress over the world ocean with error estimates. J Phys Oceanog 13:1093-1104

Hirst, AC (1989) Unstable and damped equatorial modes in simple coupled ocean-atmosphere models. J Atmos Sci 43:606-630

Kraus EB, Turner JS (1967) A one dimensional model of the seasonal thermocline. Part II Tellus 19:98-105

Levitus S (1982) Climatological atlas of the world ocean. NOAA Prof. Pap. 13, Natl. Oceanic and Atmos. Admin., Boulder, Colo.

McCreary JP, Kundu PK (1988) A numerical investigation of the Somali Current during the Southwest Monsoon. J Mar Res 46:25-58

Neelin JD, Latif M, Allaart MAF, Cane MA, Cubasch U, Gates WL, Gent PR, Ghil M, Gordon C, Lau NC, Mechoso CR, Meehl GA, Oberhuber JM, Philander SGH, Schopf PS, Sperber KR, Sterl A, Tokioka T, Tribbia J, Zebiak SE (1991) Tropical air-sea interaction in general circulation models. Climate Dyn , to appear

Pacanowski R, Philander SGH (1981) Parametrization of vertical mixing in numerical models of the tropical oceans. J Phys Oceanog 11:1443-1451

Sausen R, Barthel K, Hasselmann K (1988) Coupled ocean-atmosphere models with flux correction. Climate Dyn 2:154-163

Schopf PS, Suarez MJ (1988) Vacillations in a coupled ocean-atmosphere model. J Atmos Sci 45:549-566

Slingo A (1985) Handbook of the Meteorological Office 11-layer atmospheric general circulation model. UKMO Tech Note DCTN 29

Slingo A, Wilderspin RC, Smith R.N.B. (1989) The effect of improved physical parameterizations on simulations of cloudiness and the Earth's radiation budget. J Geophys Res 94:2281-2302

Smith RNB (1990) A scheme for predicting layer clouds and their water content in a general circulation model. Q J Roy Met Soc 116:435-460

Woodruff SD, Slutz RJ, Jenne RL, Steurer PM (1987) A Comprehensive Ocean Atmosphere Data Set. Bull Amer Met Soc 68:1239-1250

Wyrtki K, Kilonsky B (1984) Mean water and current structure during the Hawaii-to-Tahiti Shuttle Experiment. J Phys Oceanog 14:242-254

TOWARD THE GCM EL NINO SIMULATION

K. Miyakoda, A. Rosati, and R. Gudgel
U.S. Department of Commerce/NOAA
Geophysical Fluid Dynamics Laboratory
P.O. Box 308, Princeton University
Princeton, NJ 08542 USA

Abstract

The current state of El Nino simulation using air-sea coupled models is described. The first part is the excerpts from the report on Trieste Long-Range Forecast Conference in April 1991, confining only to "Simple model approach to ENSO." The second part is the report on the progress of coupled GCMs for 5 ~ 30 years integration, focusing on the El Nino phenomenon. The intense efforts on the construction of viable coupled models is being continued. A number of groups are struggling for the reproduction of El Nino events in their coupled models, expending considerable efforts and juggling various parameters relevant to the air-sea interaction. The criterion for the validity of the model is first whether the annual cycle can be simulated in the eastern Equatorial Pacific and secondly whether the ENSO signature can be produced. Overall, an appreciable amount of systematic errors have been noticed in the SST (sea surface temperature) distribution. Fortunately the light on the other side of the tunnel may be within sight. It is suggested that the appropriate surface wind stress and cumulus convection in the equatorial zone are crucial elements for satisfying the criterion. As Neelin et al. (1992) noted, the balance between the equatorial surface wind and the upwelling cold water at the eastern Pacific is critical. If one of them is excessively large or weak, the observed variability of SST would not be produced. This paper discusses a possible cause with the spectral atmospheric GCM that is responsible for generating spurious southerly trades.

NATO ASI Series, Vol. I 6
Prediction of Interannual Climate Variations
Edited by J. Shukla

1. Introduction

The objective of this paper is twofold. One is to report about the Trieste LRF (long-range forecasts) Conference, held in April 1991, confining the discussion to only air-sea coupled models for LRF. Another objective is to describe the current state of air-sea GCMs (general circulation models) for El Niño simulation studies.

It is stressed in the present paper that ultimately seasonal forecasts of weather-ocean state can be achieved by using GCMs together with appropriate initial conditions. In fact, El Niño and Southern Oscillation (ENSO) forecasts have already been semi-operationally conducted, based on several methods in which a simple dynamical model approach is included, and CAC (Climate Analysis Center) in NMC (National Meteorological Center), Washington, D.C., made an announcement of the 1991/92 El Nino at the end of November 1991 (van den Dool -- personal communication).

In order to advance the GCM seasonal forecasts further, it is pre-requisite to make a successful simulation of ENSO signature. Despite the astounding success by the simple model approach (for example, Cane and Zebiak 1985), the full fledged GCMs are still in the process of struggle, though the light at the other side of the tunnel appears to be gleaming. Several coupled models have started to exhibit interannual variabilities in the eastern Pacific, though the SST anomalies are yet weak.

2. Simple Model Approach

The ocean has a large heat capacity, and therefore it should have a good memory. Yet the feedback mechanism is not simple; the atmosphere sometimes leads the ocean in the intermonthly time-scale (Davis 1976). Thus, we here hypothesize that coupled systems could have good memories; the systems consist of the coupled ocean-atmosphere or the coupled ocean-land-atmosphere. In other words, the atmosphere itself alone is a fragile component as a memory carrier, while the coupled air-sea

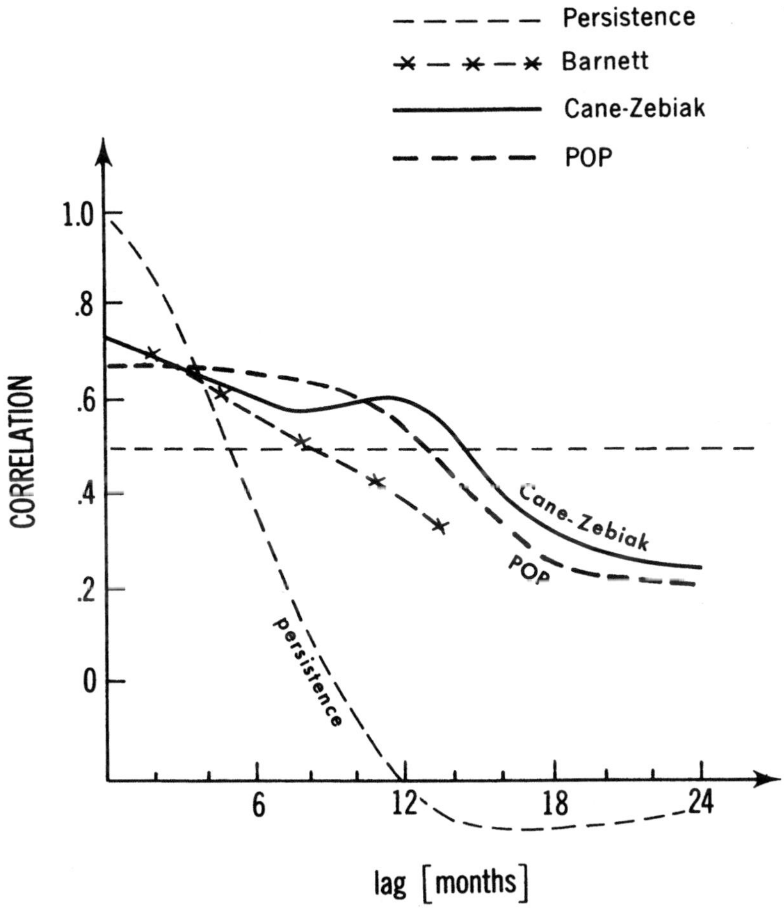

Fig. 1 Anomaly correlation curves for prediction of tropical Pacific SST (after Latif -- personal communication)

system is capable of holding prolonged, robust and distinct memory. Relevant evidences for the memory system have been presented in connection with ENSO studies (Barnett et al. 1988). The forecasts were performed either by statistical methods or by simple dynamical models (Cane et al. 1986; Inoue and O'Brien 1984). Overall forecast models capture successfully the evolution of the anomaly field of SST.

For example, in a coupled air-sea model of Cane et al. (1986), all the El Niño events (3 cases) since 1970 were predicted one or two years in advance. The forecast performance is excellent, in spite of the fact that the atmospheric model is extremely simple. It is also surprising to see that the ENSO events can be predicted as a fairly isolated system (Battisti 1989; Harrison 1989); the effect of the extratropics on ENSO is minimal, if not zero.

Thus far the first round of ENSO research has been completed (see Schopf and Suarez 1988; Philander and Lau 1988; Battisti and Hirst 1989). The ENSO memory system is now understood to consist of two parts; one is the internal waveguide of the equatorial Pacific Ocean, and the other is the atmospheric boundary layer. How far ahead in time the ENSO is predictable is a matter of argument. Goswami and Shukla (1991) investigated on the predictability of the Cane-Zebiak model (Zebiak and Cane 1987), and stated that the model has potential predictability at a time range of 9~12 months. See also Latif and Flügel (1991).

Fig. 1 displays the skill of SST (sea surface temperature) forecasts by various methods. The methods include the Cane-Zebiak model, POP method (Hamburg), in which observed heat content variation is used, based on Principal Orthogonal Pattern (POP) technique, Barnett's CCA forecasts, in which the time history of sea level pressure is used, based on Canonical Correlation Analysis (CCA) technique, and finally persistence forecast.

In parallel to these activities, the full GCM approach has been proceeded. Already some results on the seasonal forecasts started to appear. Fig. 2 (Latif et al. 1992) shows the skill of SST forecast based on a coupled GCM of MPI (Max-Planck Institute). Compared with POP method, the skill is increased. The results are based on five cases, each of which contains 4 forecast members with different initial conditions (3 months apart from each other).

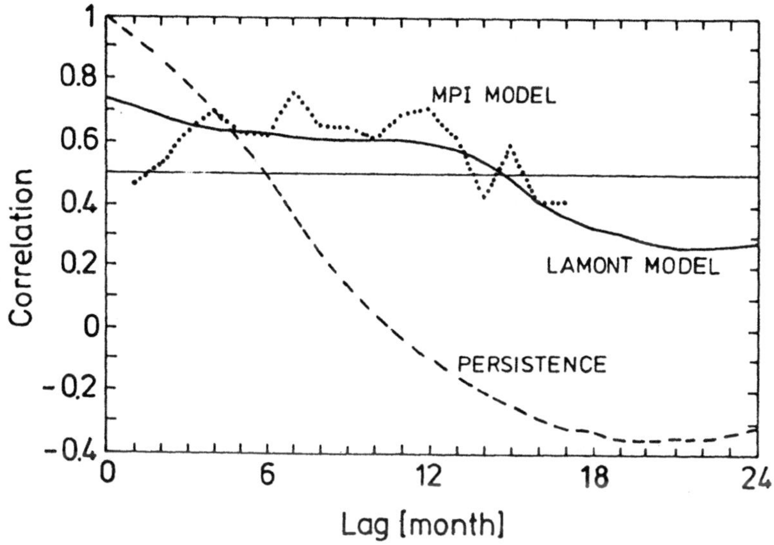

Fig. 2. Anomaly correlation curves for prediction based on coupled GCMs. Lamont -- Zebiak and Cane (1987); MPI -- Max-Planck Institute, Hamburg.(after Latif et al. 1992)

3. Current State of GCM Simulations

An extensive survey was conducted of the tropical air-sea interaction behavior of many coupled ocean-atmosphere models by Neelin et al. (1992). Here we will discuss the more current state of the extended range integrations by air-sea coupled GCMs.

i) First success

The pioneering work of Philander et al. (1992) used an air-sea GCM, the spatial resolution of which is as follows

atmos.	7.5° x 4.5° (R15)	L9	
ocean	2° x (0.3° ~ 2°)	L27	Domain: 30°S ~ 30°N

where the first column indicates the atmospheric GCM or the oceanic GCM, the second column is the horizontal resolution, the explanation being given later, the third column is the vertical resolution indicated by number of vertical layer, for example, L9 denotes 9 layers, and the fourth column is the domain used for the ocean GCM. The horizontal resolution for atmospheric model is, for example, 7.5° x 4.5° (R15), indicating that R15 (rhomboidal spectral truncation at zonal wavenumber 15) is used and this resolution model uses 7.5° longitude and 4.5° latitude Gaussian grid. On the other hand, the horizontal resolution for oceanic model is, for example, 2° x (0.3° ~ 2°), denoting that the longitudinal resolution is 2°, and the meridional resolution is 0.3° in the equatorial waveguide and is increased outside of 10°N-10°S and it is 2° at 30°N and 30°S. The oceanic domain is confined between 30°N~30°S in the Pacific Ocean, and the climatological SST is specified outside of this domain. In this model, no seasonal cycle is specified for insolation. This work has achieved for the first time in simulating the El Niño oscillations with a GCM.

ii) <u>Current state of long-runs of air-sea models</u>

There has been a notion that El Niño is an amplified event of the annual cycle of SST. In the researches described hereafter, seasonally varying insolation is forced, and the variation of SST on seasonal and interannual time scale is investigated.

UCLA				
atmos.	5° x 4°	L9		Mechoso et al. (1991)
ocean	1° x (0.3° ~ 2°)	L27	28°S~50°N	
Hamburg				
atmos.	5.6° x 5.6° (T21)	L19		Latif et al. (1992)
ocean	5° x (0.5° ~ 5°)	L17	30°S~30°N	
Hadley Centre				
atmos.	2.5° x 3.75°	L11		Davey et al. (1991)
ocean	1.5° x (0.33° ~ 1°)	L16	30°S~30°N	

MRI

atmos.	5° x 4°	L5		Nagai et al. (1992)
ocean	2.5° x 1°	L19	36°S~60°N	

MRI is Meteorological Research Institute at Tsukuba, Japan.

NMC

atmos.	3° x 3° (T40)	L18		Ji et al. (1991)
ocean	2° x (0.3° ~ 2°)	L27	30°S~30°N	

Other similar activities are under way at France, NCAR etc. In the above list, the column at right shows the reference papers. Concerning the atmospheric GCM, T21 or T40 in parentheses implies that spectral models are in use, and otherwise grid (finite difference) models are employed.

Fig. 3 is a successful example of SST along the equatorial Pacific in the longitude-time diagram (Hovmöller). All models listed above are capable of reproducing the annual variability in various degrees of the SST amplitudes in the NINO 3 region. It would be characteristic of the oceanic GCMs that many vertical levels are included, particularly resolving well for the upper 300 m depth, and that fine horizontal spacing is used in the meridional direction near the equator; the crudest one is 1° near the equator in the case of MRI.

Fig. 4 displays an example of the time variation of SST anomalies in NINO 3 region. Several GCMs realistically simulate the interannual variability, including irregular ENSO signature with preferred time scale of about 3 years (Hamburg) or 5 years (MRI). It is revealed in these experiments that the model El Niño events are locked in phase roughly to the seasonal cycle. However the magnitudes of SST anomalies are considerably small, compared with reality. Besides, not surprisingly, the simulated climatologies show sizable systematic errors.

Presently, it appears that atmospheric models with weaker trades in the eastern Pacific generate warm SST, and consequently, the coupling of such atmospheric models tend to simulate El Niño successfully. The atmospheric models which produce strong trades have considerable difficulty in generating El Niño. Neelin et al. (1992) noted that some models exhibit a weakened cold tongue along the equator, with weak zonal

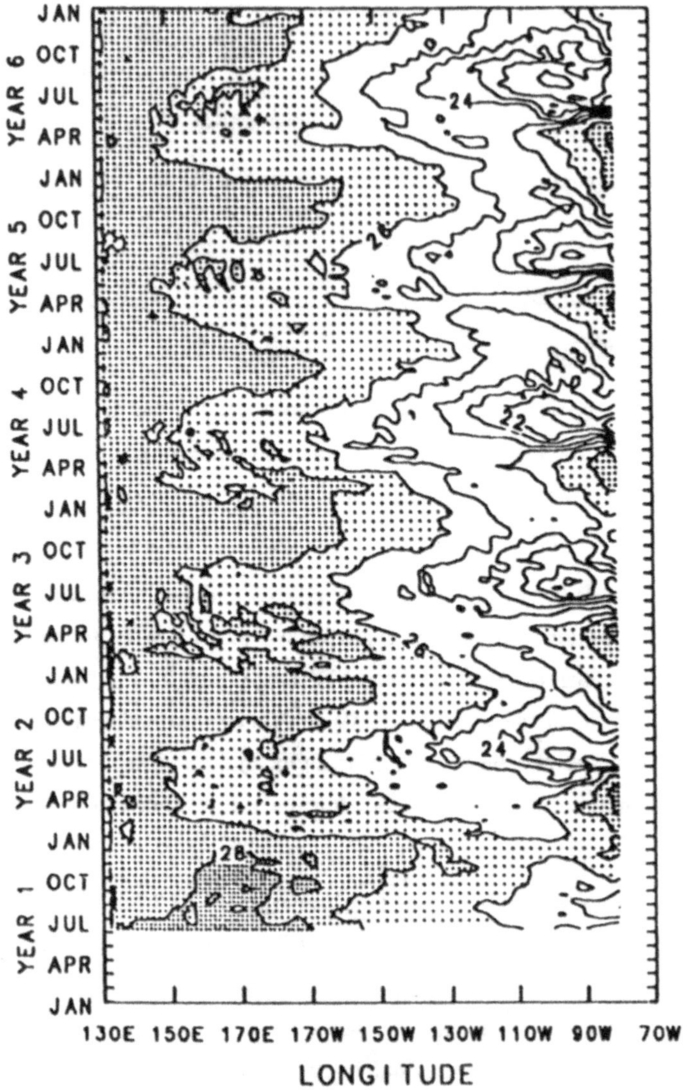

Fig. 3. Time variations of SST at equator for six years of model simulation. Contour interval is 1°C. (after Mechoso et al. 1991)

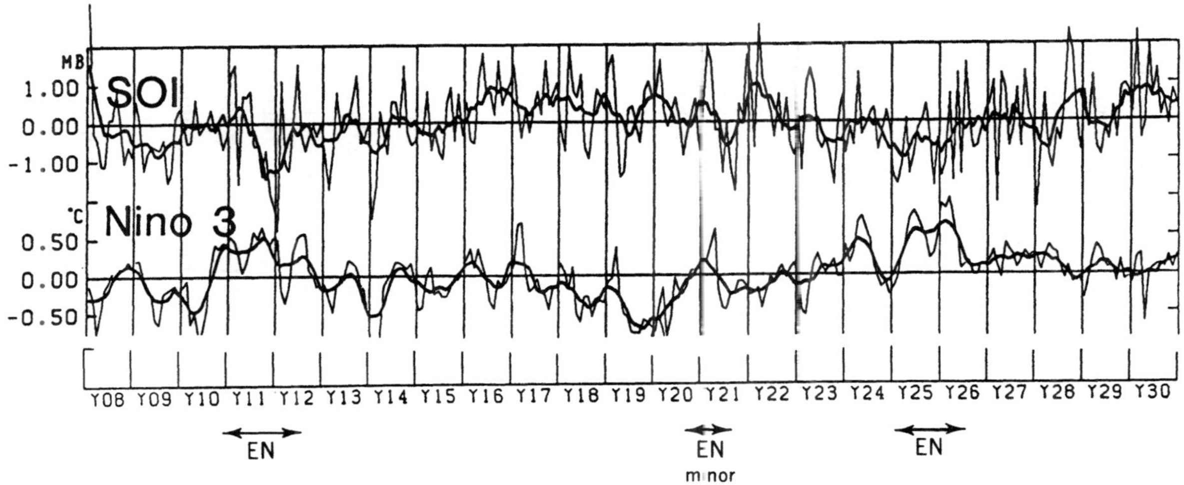

Fig. 4. Time series of model's SOI, and SST anomalies in NINO 3 by MRI model. Thin lines represent monthly mean values, while heavy line for SOI is the 13-month running mean and that for SST is the 6-month running mean.

gradients, while other models include an overly strong cold tongue. Models in which the cold tongue is permanently or seasonally weak tend not to have strong interannual variability. Neelin et al. also remarked the following. These results suggest that some of the model runs which find permanent warm conditions might give much more realistic El Niño behavior with relatively small changes in parameters. The speculation mentioned here may contain an important implication concerning the El Niño simulation. In section 5, we will come back to this point, and discuss it again with additional experience.

4. Experiments of long range simulation

The air-sea model in the Experimental Prediction Group at GFDL (hereafter referred to as the *GFDL e-model*) is as follows

atmos.	5.6° x 3.2° (R21)	L9	
	or		
	2.8° x 2.8° (T42)	L18	
ocean	1° x (0.3°~1°)	L15	Domain: global

The so-called systematic bias or the *climate drift* is fairly substantial even in the uncoupled GCM, for example, the *cold bias* in the atmospheric GCM. Trades in the eastern Pacific are excessively intense. Besides, it has been known (Sirutis and Miyakoda 1989) that this particular ocean model has a strong tendency to produce cold tongue in the eastern Equatorial Pacific. This bias is even amplified in the coupled mode.

Using this coupled model, the sensitivity test is conducted on the atmospheric model parameters. In connection with ENSO, there are dominant variables involved in this phenomenon (Fig. 5), i.e., the ocean heat content (OHC), the SST, and the surface wind stress. It is shown that these variables manifest themselves in a distinct sequence of pattern evolutions (Battisti and Hirst 1989; Philander 1990; Wakata and Sarachik

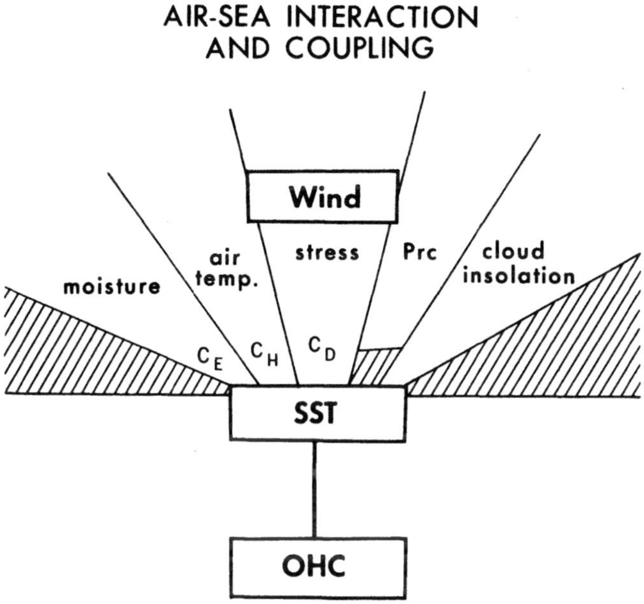

Fig. 5. Schematic illustration of air-sea interaction and coupling.

1991). Extending this concept, we consider a number of variables which affect substantially the SST evolution in the equatorial zone. First, the most basic factors directly involved in the air-sea interaction (Fig. 5) are: momentum, sensible and latent heat exchanges, which are represented by the exchange coefficients, C_D, C_H and C_E. Secondly, the factors somewhat indirectly involved in the interaction are: the incoming short-wave radiation, and the precipitation. The short-wave radiation is affected mostly by the cloudiness, and it in turn contributes to the change of SST. The precipitation is appreciably affected by westerlies/easterlies prevailing in the equatorial zone, and the condensational heating affects, in turn, the air temperature and the cloudiness.

i) Preliminary experiment

The performances of fixed cloud and interactive cloud models are compared; and the effects of evaporation and others are investigated.

The original model, *Model a* -- the fixed clouds based on the observed zonal mean clouds are used.

The modified model, *Model b* -- evaporation coefficient is reduced; the solar constant is increased; and Pacanowski's (1987) treatment of surface wind stress is used.

$$C_E \text{ (model a)} \Rightarrow 0.85 \times C_E \text{ (model b)}$$

$$\text{solar constant } I_o \text{ (model a)} \Rightarrow 1.02 \times I_o \text{ (model b)}$$

and the wind stress at the ocean surface is calculated by

$$\vec{\tau}/\rho = C_D |\vec{V}| \vec{V}$$

where

$$\vec{V} = \vec{V}_a \qquad \text{model a}$$

$$\vec{V} = \vec{V}_a - \vec{V}_o \qquad \text{model b}$$

$\vec{V}_a$ is the atmospheric surface wind vector, and $\vec{V}_o$ is the ocean surface current vector.

Presently there are several focal aspects in the air-sea coupling, i.e., reasonable level of tropical SST in NINO 4 region (western Equatorial Pacific) and annual and El Niño/La Niña variability of SST in NINO 3 region (eastern Equatorial Pacific). Figs. 6, 7 and 8 display the SST in these regions. The preliminary experiment reveals that this evaporation effect (the non-linearity in Clausius-Clapeyron equation - Newell et al. 1978) is appreciable for determining SST in the equatorial zone (*model b* > *model a*).

ii) Experiment of cloudiness

In order to investigate the effect of non-zonal cloudiness, the short wave radiation (insolation) incident to the ocean surface (SW to the ocean) is taken out, and its intensity is experimentally varied. Namely SW to the ocean is re-calculated based on the observed cloudiness or model-produced cloudiness, which is obviously non-zonal, and its intensity is increased slightly.

Model c1 and *model c2*: Fixed cloud model is used (zonal mean cloud), except that the SW to the ocean is calculated by the observed climatological cloudiness, i.e., Esbensen and Kushnir's cloudiness which varies monthly. The intensities of SW are:

$$\textit{model c1} : \textit{model c2} = 1 : 1.08$$

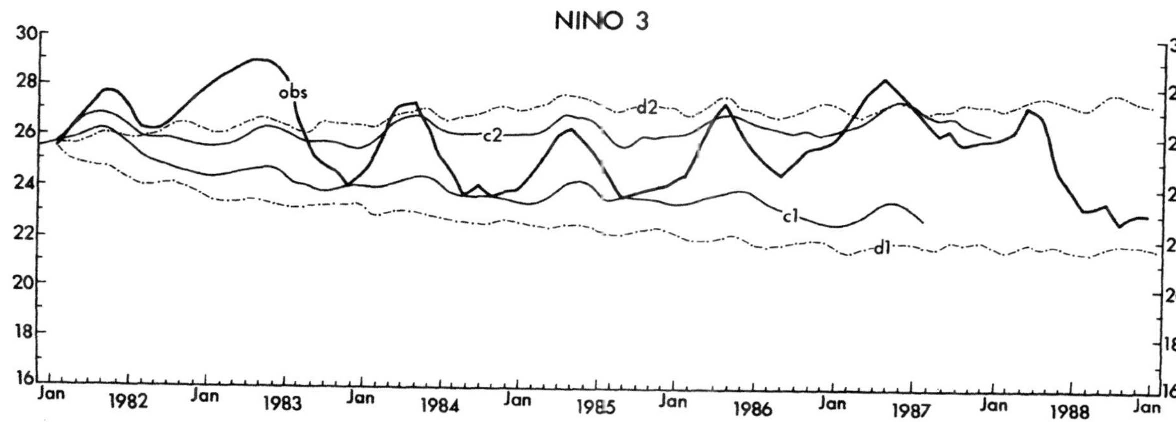

Fig. 6a. Monthly mean SST in NINO 3 region by models *c1*, *c2*, *d1* and *d2*.

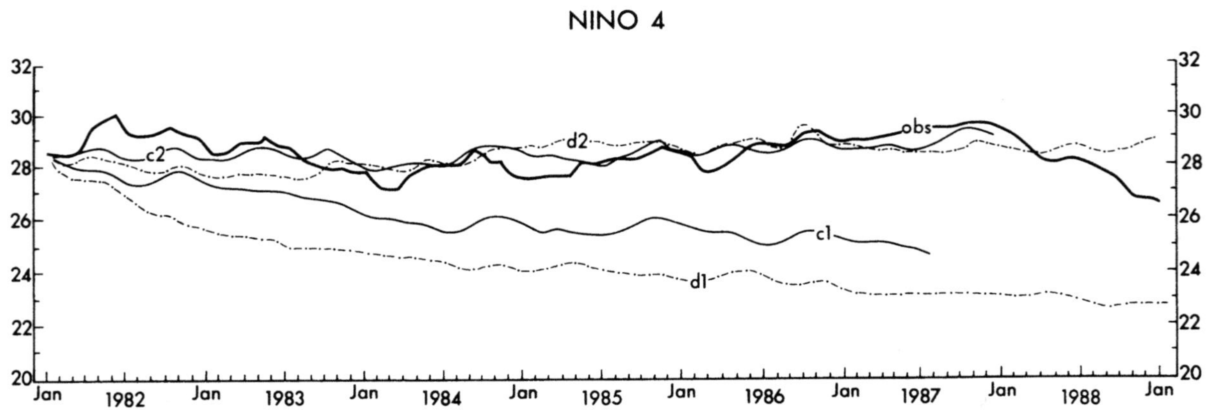

Fig. 6b. The same as Fig. 6a except for NINO 4 region.

Model d1 and model d2: These studies use the interactive cloud model (Gordon 1992). The SW to the ocean is calculated by the cloudiness in the atmospheric GCM in the same way as in *models c1* and *c2*. The intensities of SW are:

$$model\ d1 : model\ d2 = 1 : 1.15$$

Summary from Figs. 6a and b:

* Observed cloudiness (*models c1* and *c2*) gives an annual variation to SST in NINO 3, i.e., March-April with very small peak in October-November. On the other hand, the interactive cloud model (*models d1* and *d2*) gives two peaks in March-April and October-November which is not correct in the observation.
* The peaks are slightly more prominent in the observed cloud model (*models c1* and *c2*) than the interactive cloud model (*models d1* and *d2*).
* The amplitudes are slightly larger in the intensified insolation cases (*models 2* and *d2*) than in the weaker insolation cases (*models c1* and *d1*).

Yet annual variability of SST in *model c2* is substantially small compared with the observation -- implying that there must be missing factors other than the cloudiness (Figs. 6a and b).

iii) Experiment of atmospheric model resolution

In order to examine the effect of model's spatial resolution, *model d1* is used for different resolution of the atmospheric GCM, i.e., *R21L9* and *T42L18*.

As was reported elsewhere (Sirutis and Miyakoda 1989), the high resolution model generates more rainfall near the equator, and the air temperature is higher, and consequently, the SST is also higher. The results of the current experiment are shown in Figs. 7a and b. Agreeing with the previous study, there is an appreciable increase of SST in this case also, particularly in NINO 3 region.

The summary of this experiment from Figs. 7a and b is as follows.

* Overall SST level is raised in the high resolution model (*T42L18*), compared with the low resolution model (*R21L9*), particularly in the NINO 3 region.
* Annual variability of SST in NINO 3 is increased in the high resolution model. However, the amplitude of annual variation is still appreciably small, and two maxima appear erroneously every year.

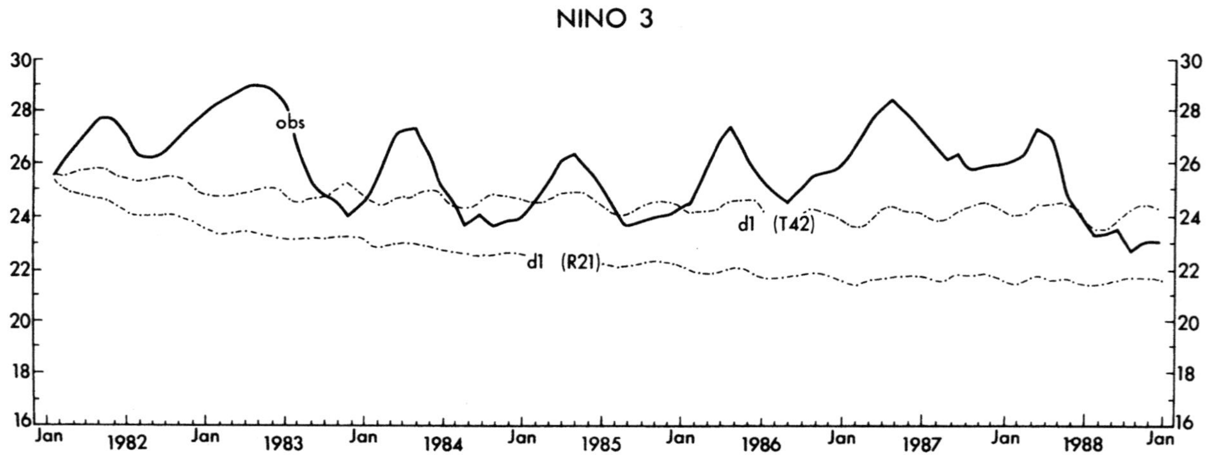

Fig. 7a. The same as Fig. 6a except by *model d1* (*R21L9*) and *model d1* (*T42L18*).

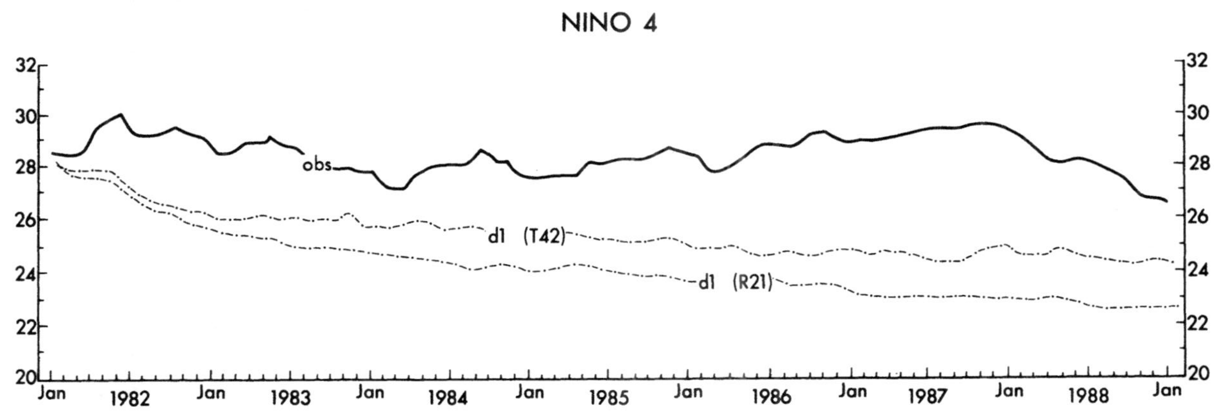

Fig. 7b. The same as Fig. 6b except by *model d1* (*R21L9*) and *model d1* (*T42L18*).

* It is interesting to note that, despite the marginal increase of SST with *model d1-T42* in NINO 4 (Fig. 7b), SST map shows a tendency to form a pronounced warm pool in the western Pacific. The tropical SST distributions over Atlantic and Indian Oceans are noticeably improved in *model T42*, compared with observation. Yet the great deficiency is apparent in the ITCZ (Inter-tropical Convergence Zone) near Mexico.

iv) Experiment of wind stress

μ-run (data assimilation mode)

Deviating slightly from the line of simulation experiment, the effect of surface wind is investigated by applying the observed wind data to the coupled model run. This is a simplified data assimilation, based on Newtonian nudging technique. Taking the coupled air-sea model which we have used in the preceding sections, two equations are modified. The modification is made only for the momentum in the lowest layer of the atmospheric GCM, i.e.,

$$\frac{\partial \vec{V}}{\partial t} \cdots\cdots = -\mu\,(\vec{V} - \vec{V}_{obs}) \qquad (4.7)$$

where $\vec{V}$ is the wind vector, the term $-\mu\,(\vec{V} - \vec{V}_{obs})$ is the Newtonian nudging term, μ is the damping coefficient, $(1\ \mathrm{hr})^{-1}$, $\vec{V}_{obs}$ is the observed wind vector. In this study the NMC analysis of 12 hourly wind data at 1000 hPa level are utilized.

The SST curves in Figs. 8a and b, labeled by *μ-run*, are the results of this approach, in which the basic coupled *model d2* of *R21L9* is used. As was speculated, first SSTs in both regions are increased substantially, due to the better specification of trade winds; and secondly the annual variability as well as El Niño signature are appreciably enhanced. Admittedly, however, the quality of NMC analysis was not high before 1986 data set (Trenberth and Olson 1988). In addition, *model d2* in this particular version (*d2* = 1.15 *d1* in SW) gives excessively high temperature at the equatorial zone.

The synoptic map of SST reveals that non-symmetric feature of SST distribution has now been produced by the observed wind stress. It is indeed intriguing to see how large effect is caused by the slight modification of wind fields. According to Gutzler and Wood (1990) and Nigam and Shen (1992), the winds crucial for the cumulus convection east of the dateline are the meridional components, while those for west of the dateline are the zonal components. Furthermore, the convergence crucial for the cumulus activities east of the dateline is due to the surface winds, whereas that for west of the dateline is due to the 850 hPa winds.

The climatological character of the uncoupled atmospheric GCM (*GFDL e-model*) is investigated based on the 10 year run with a specified observed SST (Stern and Miyakoda 1991). The systematic bias was estimated by using the ECMWF (European Centre for Medium-Range Weather Forecasts) analysis as the reference. This study reveals that (a) in the eastern Pacific there is an extraordinarily large bias in meridional winds, the *T42L18* model produces negative bias of as much as 4 ms^{-1} at 850 hPa level, i.e., weaker southerlies; (b) in the equatorial zone there is a bias of the zonal winds, the model produces negative bias of 3 ms^{-1} at 150°W longitude, i.e., stronger easterlies, though it produces positive bias of 4 ms^{-1} at 150°E. In summary, the weaker southerlies in the eastern Pacific is indeed the fatal cause for the bias, which can sufficiently explain most of the drawbacks in the air-sea model.

Forecasting with surface-wind-correction

Based on the information in this data assimilation study, another experiment is conducted with *model e2* of *R21L9*. First the output of *μ-run* is averaged with time; the quantity in concern is

$$\tau_d = -\int \mu (\vec{V} - \vec{V}_{obs}) \frac{\Delta p}{g} \, dt \, / \int dt \tag{4.8}$$

where Δp is the thickness of the lowest layer of the atmospheric GCM. The quantity τ_d is stored for some years with monthly stratification. In *model e2*, the statistically averaged value of τ_d is utilized as follows

$$\frac{\partial \vec{V}}{\partial t} \cdots\cdots = \tau_d / \left(\frac{\Delta p}{g}\right)$$

The SW to the ocean is (see (4.6))

$$model\ d1 : model\ e2 = 1 : 1.075 \tag{4.9}$$

The results of *model e2* are shown in Figs. 8a and b. The annual variation of SST in NINO 3 is substantially improved; it has one peak in March-April and one minimum in July-August, though this feature was not established until the third year. The

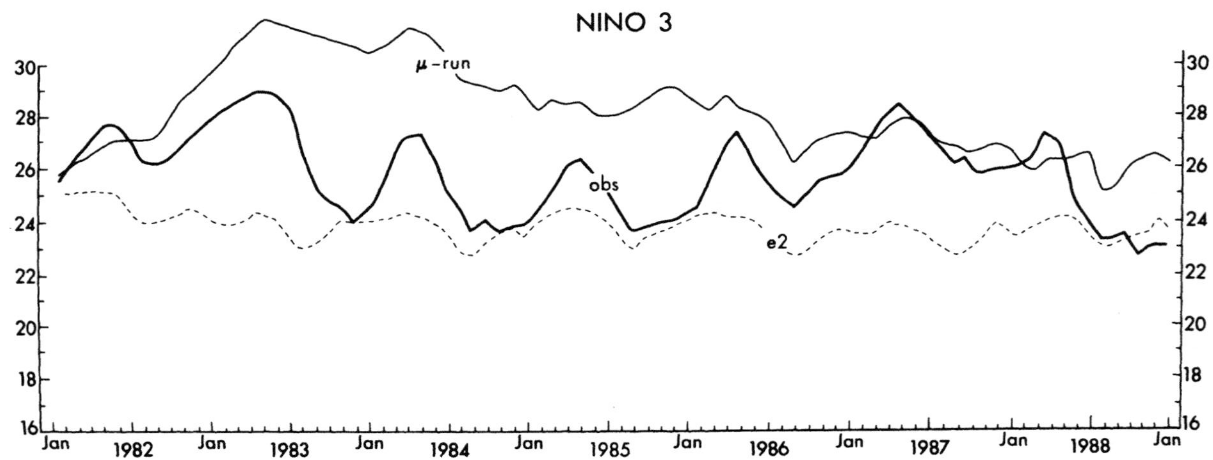

Fig. 8a. The same as Fig. 6a except by μ-*run* and *model e2*.

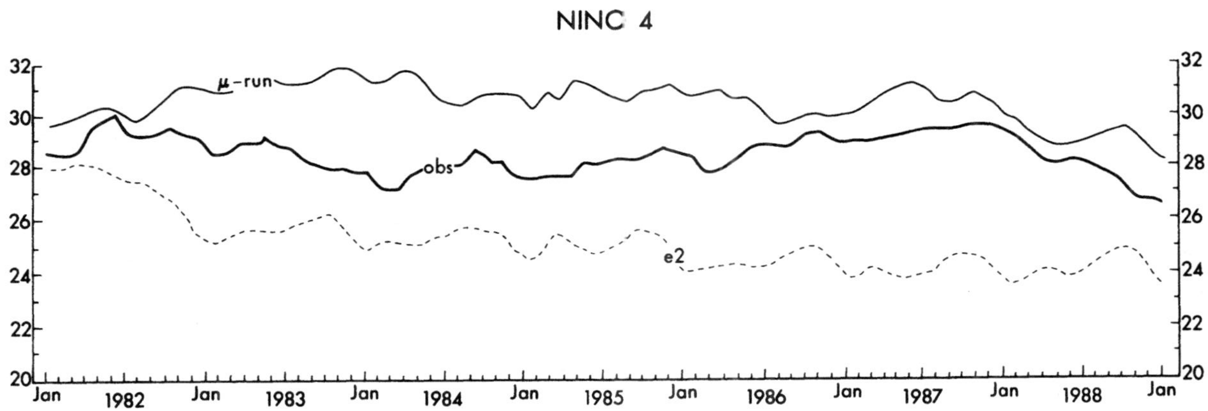

Fig. 8b. The same as Fig. 6b except by μ-*run* and *model e2*.

simulation has now been extended to 15 years, and the results reveal clearly the interannual variability. However, the amplitude of the variation is still not large. It is expected, though, that this amplitude is increased in the *T42* model, and more importantly, that the improved cumulus parameterization may enhance the El Nino signature.

5. Remarks

The center of the weaker southerly trade bias is located at the coast of Chili, and the error in the *GFDL e-model* is produced as the Gibbs phenomenon of the spectral model, associated with the Andes. Spectral models have often difficulty to handle steep mountain. On the other hand, UCLA, Hadley Centre, and MRI models use the finite difference scheme, presumably avoiding this difficulty. Note, however, that NMC and Hamburg use spectral models, and yet they are successful in obtaining the annual cycles. Possibilities are that Gibbs' problems are avoided by using less steeper Andes (Hamburg, see Latif et al. 1992) or that their careful arrangement of cumulus, boundary layer and interactive cloud parameterizations manage to avoid the difficulty (NMC, see Ji et al. 1991).

Secondly, the cumulus parameterizations in UCLA, MRI, Hadley Centre, and Hamburg models are totally different from that of GFDL; the former two employ the Arakawa-Schubert method, and Hadley Centre and Hamburg use Rowntree's scheme and Kuo's scheme, respectively, while the *GFDL e-model* uses the "moist convective adjustment". There has been a view that the intensity of cumulus activities near Philippine, New Guinea and central Pacific is vital for warming the western and central Pacific, and that these cumulus updrafts control the Walker Circulation, which in turn affects the trade winds. Another possibility is that the cumulus activities over the ITCZ near Mexico are also critical for generating adequate meridional components of surface winds. In the Hadley Centre's case, their coupled model started to exhibit the SST annual cycle in NINO 3 only after introducing a parameterization of "gustiness", which contributes to the increase of evaporation in the convective region and eventually to reduce SST.

However, even if the surface boundary conditions for the ocean model were perfect there could exist a systematic error in the ocean model that produces unrealistically strong upwelling in the eastern Equatorial Pacific. This bias would contribute to the cold tongue being too cold and extending too far west. This could be due to the "inhomogeneous and non-isotropic grid structure" of ocean GCM.

Collecting the lessons from section 4, we summarize the essence about the GCM simulation of interannual variability. As mentioned in section 3, the atmospheric models with weaker trades in the eastern Pacific generate warm SST, and as a result, they tend to provide much higher probability of a successful simulation of El Niño. More precisely speaking, stronger southerly and weaker easterly trades in the eastern tropical Pacific are favorable.

The path toward the success of El Niño simulation does not appear wide. Besides, at the present stage, models in general contain an appreciable amount of climate bias. Therefore, a special strategy is needed for the construction of viable coupled models. It is suggested first to make an intentional effort for simulating El Niño. Intentional efforts mean utmost attempts with use of various techniques, which are normally not recommendable, such as the heat flux correction, the surface wind correction, juggling of the coefficients of evaporation, modulating of cumulus parameterization, introduction of "gustiness", adjustment of penetration of insolation into the ocean, smoothing Andes, etc., though some techniques are counter-productive in some situations (see Latif et al. 1992).

After managing to achieve the El Niño reproduction, one can gradually reduce the systematic bias by increasing the models resolution, improving the physics in reasonable and acceptable ways, and removing tentative tricks such as "flux correction" and others.

6. Conclusions and Comments

First the Trieste Long-range Forecast Conference in April 1991 is reported, limiting the discussion to only the simple model approaches to seasonal forecasts. Secondly, the current state of the El Niño simulation studies using air-sea coupled models is described. In the latter, some on-going studies with the GFDL model (Experimental Prediction Group) are included.

Tentative conclusions:

Air-sea coupled models in some groups have been successful in generating the annual variability of SST in the NINO 3 region as well as the El Niño signature (Hamburg and MRI), though SST anomalies are yet weak. Which factor is most responsible for the success in simulation of El Niño processes is a major concern in this paper. It is suggested that the adequate surface wind stress in the eastern Equatorial Pacific and the intense cumulus activities in the western and central equatorial zone appear essential.

The fatal deficiency of the GFDL model has been due to the failure in generating proper meridional surface winds at the eastern Equatorial Pacific. It is suggested that this deficiency is mainly caused by the spectral Gibbs phenomenon, associated with the steep Andes orography. However, the annual variability of SST in NINO 3 region can be produced by "surface-wind-correction" method. On the other hand, the weaker convergence bias in the 150°W equatorial zone is another drawback, which would be related to the cumulus parameterization.

The grid resolution of atmospheric GCM is an important factor for successful simulation of El Niño phenomenon. The experience of using *T42L18* model suggests that the atmospheric GCM is desired to have even higher resolution.

Comments:

* The simulation of surface winds has to be improved substantially. Otherwise "surface-wind-correction" method has to be applied.

* The ocean GCM of GFDL model has a fine grid resolution in the equatorial zone so as to represent the ocean internal waves. This constraint has forced to make ocean

grid structure unhomogeneous and non-isotropic. A homogeneous and isotropic grid structure of ocean GCM would be more desirable, i.e., a global ocean model of 0.5° x 0.5° resolution.

References

Barnett T P, Graham N, Cane M, Zebiak S, Dolan S, O'Brien J, Legler D (1988) On prediction of the El Nino of 1986-87. Science 241:192-196

Battisti D S (1989) On the role of off-equatorial oceanic Rossby waves during ENSO. J Phys Oceanogr 19:551-559

Battisti D S, Hirst A C (1989) Interannual variability in a tropical atmosphere-ocean model: Influence of the basic state, ocean geometry and nonlinearity. J Atmos Sci 46:1687-1712

Busalacchi A J, O'Brien J J (1981) Interannual variability of the equatorial Pacific in the 1960's. J Geophys Res 86:10901-10907

Cane M A, Zebiak S E (1985) A theory for El Niño and the Southern Oscillation. Science 228:1084-1087

Cane M A, Zebiak S E, Dolan S C (1986) Experimental forecasts of El Niño. Nature 321:827-832

Davey M K, Gordon C, Ineson S, Lawrence S (1991) Coupled tropical ocean global atmosphere models at the UKMO. In: Climate Research Technical Note No. 21:Hadley Centre for Climate Research and Prediction UK, p 17

Davis R E (1976) Predictability of sea surface temperature and sea level pressure anomalies over the North Pacific Ocean. J Phys Oceanogr 6:249-266

Gutzler D S, and Wood T M (1990) Structure of large-scale convective anomalies over tropical oceans. J Climate 3:483-496

Gordon C T (to be published) Comparison of 30 day integration with and without cloud-radiation interaction. Mon Wea Rev

Goswami B N, Shukla J (1991) Predictability of a coupled ocean-atmosphere model. J Climate 4:3-22

Harrison D E (1989) Local and remote forcing of ENSO ocean waveguide response. J Phys Oceanogr 19:691-695

Inoue M, O'Brien J J (1984) A forecasting model for the onset of a major El Niño. Mon Wea Rev 112:2326-2337

Ji M, Kumar A, Leetmaa A (1991) Coupled interaction in the equatorial eastern Pacific. In: Proceedings of the Sixteenth Annual Climate Dynamics Workshop held at UCLA CA:28 October-1 November 1991, p 144-152

Latif M, Flügel M (1991) An investigation of short range climate predictability in the tropical Pacific. J Geophys Res 96(C2):2661-2673

Latif M, Sterl A, Maier-Reimer E, Junge M M (to be published) Structure and predictability of the El Niño/Southern Oscillation phenomenon. J Climate

Mechoso C R, Ma C-C, Farrara J D, Spahr J A, Moore R W (1991) Distribution of a climate model across high-speed networks. In: Proceedings of the Supercomputing '91. Albuquerque NM 18-22 November 1991:IEEE Computer Society p 253-260

Nagai T, Tokioka T, Endoh M, Kitamura Y (to be published) El Niño/Southern Oscillation simulated in an MRI atmosphere-ocean coupled general circulation model. J Climate

Neelin J D et al (to be published) Tropical air-sea interaction in general circulation models. Clim Dyn

Newell, R E, Navato A R, Hsiung J (1978) Long-term global sea surface temperature fluctuations and their possible influence on atmospheric CO_2 concentration. J Pure and Appl Geophys 116:351-371

Nigam S, Shen H-S (to be published) Structure of oceanic and atmospheric low-frequency variability over the tropical Pacific and Indian Oceans. Part I: COADS observations. J Climate

Pacanowski R C (1987) Effect of equatorial currents on surface stress. J Phys Oceanogr 17:833-858

Philander S G H (1990) A review of simulations of the Southern Oscillation. In: International TOGA Scientific Conference Proceedings held at Honolulu, HI:16-20 July 1990. World Climate Research Programme WCRP-43 WMO/TD No. 379

Philander S G H, Pacanowski R C, Lau N-C, Nath M J (1992) Simulation of ENSO with a global atmospheric GCM coupled to a high-resolution, tropical Pacific ocean GCM. J Climate 5:308-329

Philander S G H, Lau N-C (1988) Predictability of El Niño. In: Schlesinger M E (ed) Kluwer Academic Publishers, Physically Based Modelling and Simulation of Climate and Climate Change Part II: p 967-982

Schopf P S, Suarez M J (1988) Vacillations in a coupled ocean-atmosphere model. J Atmos Sci 45:549-566

Sirutis J, Miyakoda K (1989) Systematic bias in predictions with a coupled air-sea model. In: Proceedings from the Workshop on Systematic Errors in Models of the Atmosphere held at Toronto, Canada. September 1988. WMO/TD No. 273 p 136-140

Stern W F, Miyakoda K (1991) The feasibility of seasonal atmospheric prediction speculated from SST forced decadal simulations. In: Proceedings of Ninth Conference on Numerical Weather Prediction held at Denver CO. 14-18 October 1991:Amer Meteor Soc p 649-651

Trenberth K E, Olson J G (1988) Evaluation of NMC global analysis 1979-1987. In: NCAR Tech Note, NCAR/TN-299+STR, NCAR Boulder CO, p 82

Wakata Y, Sarachik S E (1991) On the role of equatorial ocean modes in the ENSO cycle. J Phys Oceanogr 21:434-443

Zebiak S E, Cane M A (1987) A model El Niño-Southern Oscillation. Mon Wea Rev 115:2262-2278

INTERANNUAL CLIMATE VARIATIONS OVER THE TROPICAL ATLANTIC OCEAN

J. Servain
Centre National de la Recherche Scientifique
Laboratoire d'Océanographie Dynamique et de Climatologie
Centre ORSTOM (IFREMER)
29280 Plouzané (France)

J. Merle
Institut Français de Recherche Scientifique pour le Développement en Coopération
Laboratoire d'Océanographie Dynamique et de Climatologie
Université P. et M. Curie
75252 Paris Cédex 05 (France)

1. INTRODUCTION

The capability to predict climate variations on various time scales depends on the ability to comprehend the complex interplay of many processes which determine the present climate and its variability. Through the years, various scientists have correlated the low-frequency atmospheric oscillations to changes occurring in the tropical oceans. The principal elements of the tightly coupled tropical atmosphere-ocean system are the relatively fast dynamical response of tropical ocean basins to changes in the surface winds, associated changes in mixed layer depth, heat content and surface temperature (SST) and, in the atmosphere, the response of tropical convection, precipitation and surface winds to variation of SST.
The purpose of the current 10-year program Tropical Ocean - Global Atmosphere (TOGA) is to observe and understand year-to-year climate variability. The El Niño-Southern Oscillation (ENSO) phenomenon, with its marine effect which occurs principally over the eastern equatorial Pacific Ocean, is undoubtedly the most noted and pronounced case of such variability. ENSO events, as for instance that developed during 1982-1983 (Quiroz, 1983), received justifiably explosive interest during the last decade that somewhat shadowed the role and importance of other ocean sectors.
Despite its smaller size and probably much weaker influence in the low latitude air-sea interaction system compared with the Indonesian-Pacific oceanic sector, the Atlantic Ocean is affected by a large spectrum of climatic signals (Hastenrath & Kaczmarczysk,

NATO ASI Series, Vol. I 6
Prediction of Interannual Climate Variations
Edited by J. Shukla

1981). Furthermore, the Atlantic Ocean is a meridional ocean and its tropical part is one of the principal meridional heat exchange region of the world's Ocean (Stommel, 1979) and so has an important influence on the global earth's climate at large time scales.
What is intended here is to clarify the relation between the tropical Atlantic Ocean variability and other dominant climatic signals. In particular, we investigate how the interannual signal of the Atlantic is leaded by ENSO, or is controled by its own internal mode of variability, including the North Atlantic Oscillation (NAO).

2. SEASONAL AND INTERANNUAL VARIABILITY OF THE TROPICAL ATLANTIC OCEAN

The tropical Atlantic Ocean is in many respects similar to the tropical Pacific Ocean, having for example, the same wind structure. Northeast and Southwest trade winds, converging in a near-equatorial Intertropical Convergence Zone (ITCZ) with similar seasonal variations, induce almost identical large scale circulation systems (Fig 1). Regions of maximum seasonal variability of the wind stress magnitude straddle the mean position of the ITCZ whereas the largest deviations of the wind stress direction occur along this line.
Domains of significant interannual wind stress fluctuations do not necessarily coincide with the domains of maximum seasonal variability (Servain et al., 1985). The amplitude of the interannual variability of the meridional wind stress is nearly constant over broad regions of the tropical basin in reverse to the amplitude of the zonal wind stress which is reduced in the Gulf of Guinea.
The ratio of the amplitude of the seasonal cycle to the amplitude of interannual perturbations is similar to that of the seasonal variability. Monthly anomalies of the wind stress are greater than the amplitude of the seasonal variability for large regions of the tropical Atlantic including the western equatorial zone. The seasonal wind stress changes are dominant only in the vicinity of the ITCZ.
Based on the analyses of SST data in the Gulf of Guinea by Merle et al. (1980) and theoretical consideration of the adjustement time scale of the basin (Philander, 1979), it is asserted that the tropical Atlantic thermal variability is dominated by a large

seasonal cycle. The shallowing thermocline in the eastern tropical Atlantic together with vertical mixing induces maximum seasonal SST variability in the seasonal upwelling zones along the West African coast and along the equator near 10°W (Fig 2). The location of the minimum SST variability coincides with the thermal equator and hence the mean ITCZ position.

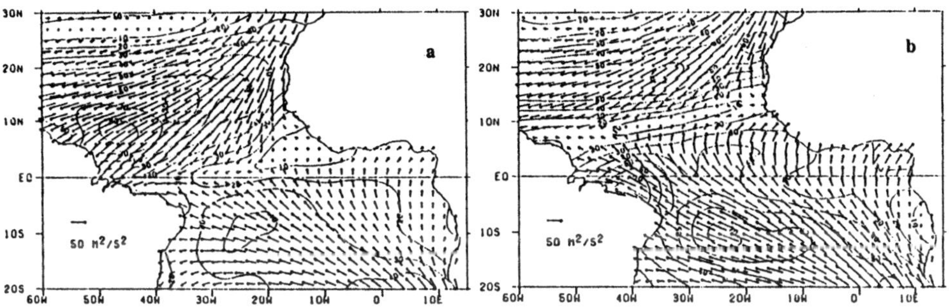

Figure 1 : (a) Climatic monthly average of pseudo wind stress for February and (b) August [adapted after Picaut et al., 1985]

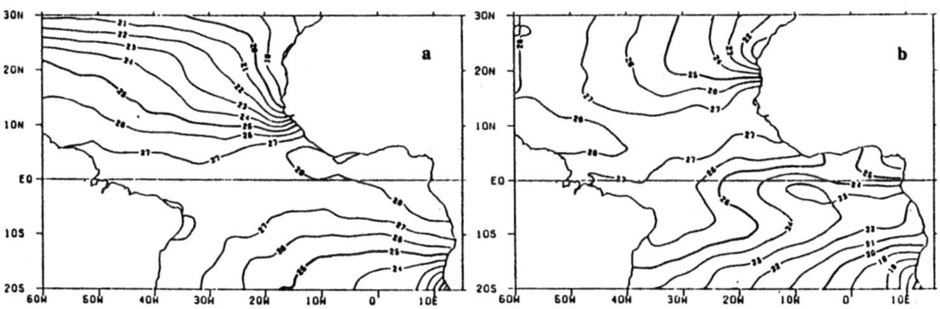

Figure 2 : (a) Climatic monthly average of SST for February and (b) August [adapted from Picaut et al., 1985]

The interannual variability of SST is maximum in the coastal and equatorial regions (Servain et al., 1985). High values of the ratio of the seasonal variability to the interannual variability are found in the regions where the annual signal is important. For the western and central equatorial Atlantic Ocean this ratio is less than two and is less than one in the vicinity of the thermal equator.

From the analysis of long-term time series of anomalies of wind stress and SST, it has been shown (Servain et al., 1985; Servain &

Legler, 1986) that anomalous events can persist for more than one year in the tropical Atlantic Ocean. Anomalies of such duration cannot be the result of a simple phase shift or change in amplitude of the seasonal cycle. Of particular importance are the wind stress anomalies of the western equatorial Atlantic that may be capable of inducing a dynamical response in the Gulf of Guinea greater than or equal to the response to the mean seasonal forcing (Busalacchi & Picaut, 1983; Servain, 1984).

The most spectacular example of long-term and large amplitude of interannual SST anomalie occurred in the tropical Atlantic about one year after the 1982-1983 Pacific EL Niño: a warm event of unusual amplitude (SST anomalies over 3.5 °C) has been observed in the tropical Atlantic in June-September 1984 following a relaxation of the zonal wind stress in the first six months of 1984, after the 1982-1983 El Niño of the Pacific. During the first half of 1984, oceanic and atmospheric conditions in the tropical Atlantic were, in many respects, similar to conditions in the Pacific Ocean during El Niño. The higher SSTs in 1984 were related with exceptionally heavy rainfall over Brazilian Nordeste and the coastal zone of southwestern Africa but were also associated with a persistence of the drought in the Sahel.

The ITCZ position varies seasonally over a large latitudinal band which extends roughly from the equator to 10-15° North (Fig 1). By contrast to that occurs with the zone of maximum SST (Servain, 1991), the meridional shift of ITCZ is generally a little larger in the eastern than in the the western part of the basin. At 28°W its northern and southern limits are reached at the end of each extreme season, that is, in March and August respectively (Fig 3). Anomalous meridional displacements of this kinematic axis separating northern and southern hemisphere trades are in relation with an asymmetrical thermal structure compared to the equator (Lamb, 1978a; Servain, 1985) : a more southward (northward) position of the axis is associated with warmer (cooler) conditions in the southern part of the basin and cooler (warmer) conditions in the northern part (Fig 4). This meridional dipole mode of interannual Atlantic variability of SST and ITCZ position is related with rainfall anomalies in both the African Sahel region (e.g., Lamb, 1978a,b) and the Brazilian Nordeste region (e.g., Hastenrath & Heller, 1977; Moura & Shukla, 1981), (Table 1).

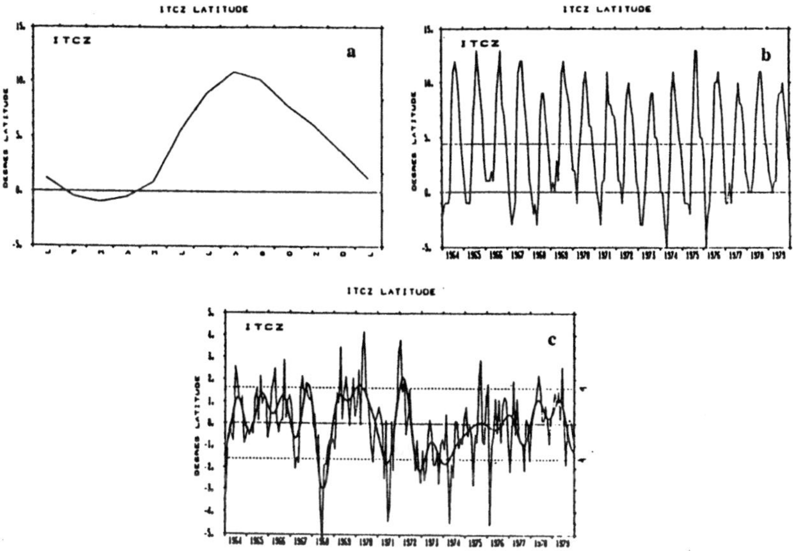

Figure 3 : Time series of the latitudinal position of ITCZ along the 28°W meridian computed during the years 1964-1979: (a) climatic monthly mean values, (b) interannual monthly mean values, and (c) interannual monthly anomalies [after Servain, 1985]

Table 1 . Correlation Coefficients Between Calendar Month Values of North Tropical Basin (NB), South Tropical Basin (SB), and Dipole (NB-SB), with the 1964-1987 NORDESTE Index, and the 1964-1988 SAHEL Index [adapted after Servain, 1991]

	NORDESTE	SAHEL
NB (FEB)	-0.44	
SB (FEB)	0.38	
NB (FEB) -SB (FEB)	-0.65*	
NB (AUG)		0.14
SB (AUG)		-0.40
NB (AUG) -SB (AUG)		0.46

NB and SB domains are the same as in Figure 16. The calendar month is February (FEB) or August (AUG). NORDESTE and SAHEL indices have been provided by S. Hastenrath and P. Lamb (person. comm., 1991 and 1988 respectively).The bold values are significant at the 5% level.
* Significant at 1% level

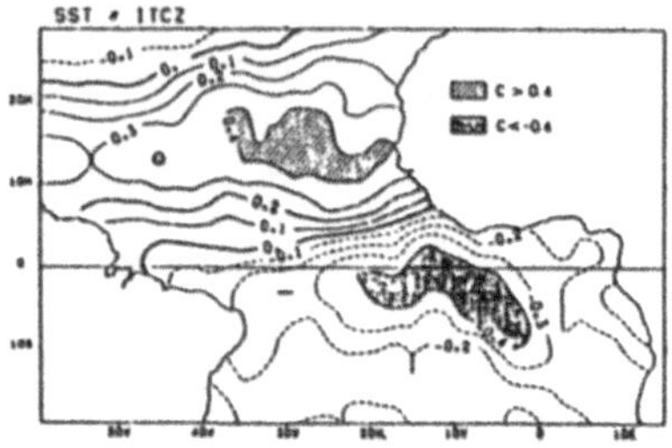

Figure 4 : Correlation coefficient pattern at zero lag, computed during the years 1964-1979, between the time series of the abnormal latitudinal position of ITCZ at 28°W and the monthly anomalies of SST inside the tropical Atlantic Ocean [after Servain, 1985]

The 1984 Atlantic warm event was also associated with an abnormal position of the ITCZ. The narrow east-west band of rising air, cloudiness and heavy rainfall onto the convergence zone between the southeast and northeast trades was displaced unusually far southward during 1984. Such a displacement of the ITCZ is also a feature of El Niño in the Pacific Ocean.

3. RELATIONS WITH ENSO

The equatorial Atlantic sea level pressure (SLP) field is under the influence of the "Darwin pole" and represents a secondary region of high correlation of the SO of opposite sign with the Tahiti pole (Fig 5). This Atlantic secondary center of action was already put in evidence by Kidson (1975). Seasonal variations of the correlation between the tropical Atlantic SLP and Darwin SLP exist and are found the greatest during the boreal winter and in the western part of the equatorial Atlantic basin (Tourre, person. comm., 1988).
Although the variance of ITCZ position over the Atlantic is smaller than of the Pacific, the interannual variabilty of their "intensities" - such as described by the observed outgoing longwave radiation (OLR) anomalies (Fig 6) - are roughly of the same range (Daniels, 1985). The inverse relationship between the equatorial Pacific and both Indonesia and Atlantic equatorial regions is evidenced during major ENSOs phenomena, as for example that one of years 1982-1983.

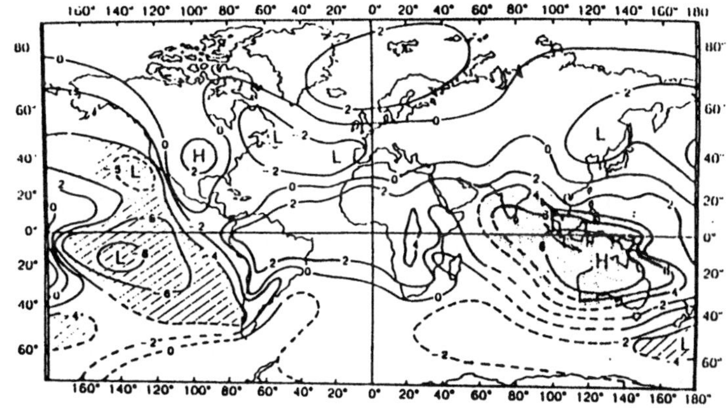

Figure 5 : Pattern of correlation coefficients between the global annual mean SLP with Darwin SLP [after Trenberth & Shea, 1987, adapted by Tourre, person. comm., 1988]

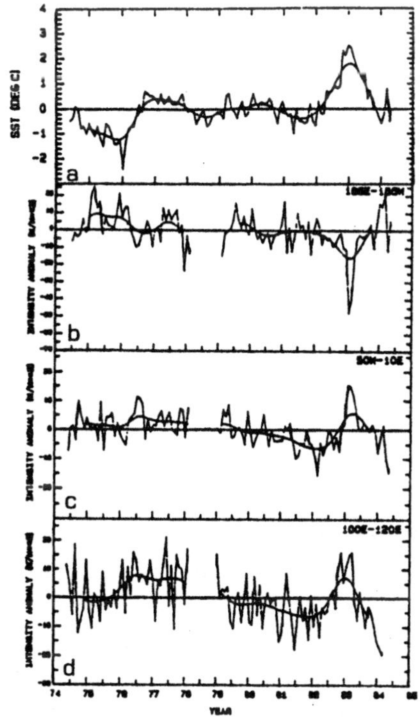

Figure 6 : (a) Time series of monthly mean SST anomaly for the Key region at 130W, (b) equatorial Pacific OLR anomaly averaged between 165E-165W, (c) equatorial Atlantic OLR anomaly averaged between 50W-10E and (d) equatorial Indonesian region OLR anomaly averaged between 100E-120E, for the period June 1974 - April 1984 [after Daniels, 1985]

A number of studies have shown that changes in atmospheric conditions over the tropical Atlantic Ocean seem to follow (by a few months) the changes that occur in the Pacific (Tourre, person. comm., 1988). Early in 1982, SSTs in the eastern Pacific rised rapidly. Simultaneous with this increase were a decrease of the surface trade winds and an increase of ITCZ intensity in eastern and central Pacific as indicated by negative OLR fluctuations (Fig 6). Horel et al. (1986) documented a strong abnormal Walker-type circulation consequently emphased between the eastern equatorial Pacific and the equatorial Atlantic region (Fig 7). From mid-1982 to mid-1983 above the equatorial Atlantic, the westerly flux at 20 kPa increased while the trade winds at the sea surface remained stronger than normal. During that time, in the same region, there was a decrease of convective activity which was minimum in February 1983 (Fig 6). A reversal situation occurred drastically from mid-1983 until beginning of year 1984 with relaxations of zonal winds at both 20 kPa surface and sea surface, accompanied with a peak of convective activity over the equatorial Atlantic. This important relaxation of the wind stress (Fig 8) was associated with an eastward motion of the center of the St. Helena Anticyclone. As a consequence the upwelling all along the west coast of Africa was suppressed, inducing the largest SST anomalies ever recorded in the Gulf of Guinea (Servain & Séva, 1987). This dramatic Atlantic warm event was associated with the inverse tilting of the dynamic equatorial ocean slope observed during the FOCAL oceanographic experiments in 1984 (Katz et al., 1986).

The highest correlation coefficients between ENSO index and the tropical Atlantic Ocean SST situation are found using the Atlantic northern part when ENSO leads it by a few months (Servain, 1991). To illustrate that, Figure 9 provides two opposite SST anomaly fields observed in the tropical Atlantic : the first one (strongly negative in the north) is the pattern in March-April 1974, a few months after the most extreme anti-El Niño recorded during the last three decades, and the second one (strongly positive in the north) is the pattern in March-April 1983, just after the peak of the dramatic 1982-1983 El Niño. In fact, it seems that the ENSO phenomenon correlates well with a lagged Atlantic dipole index (South basin

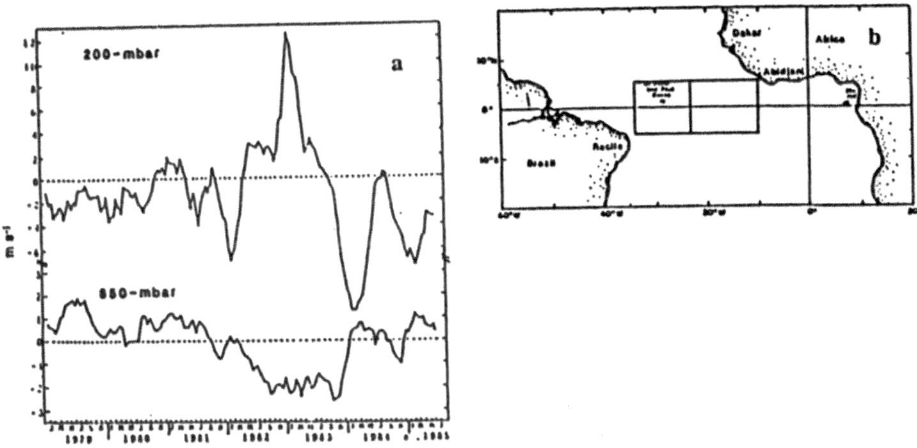

Figure 7 : (a) Indices of 200-mbar zonal wind anomalies (top) and 850-mbar zonal wind anomalies (bottom). The indices are based on observations within the region 5N to 5S and from 35W to 10W (b). The anomalies are computed with respect to the 1978-1983 base period [after Horel et al., 1986]

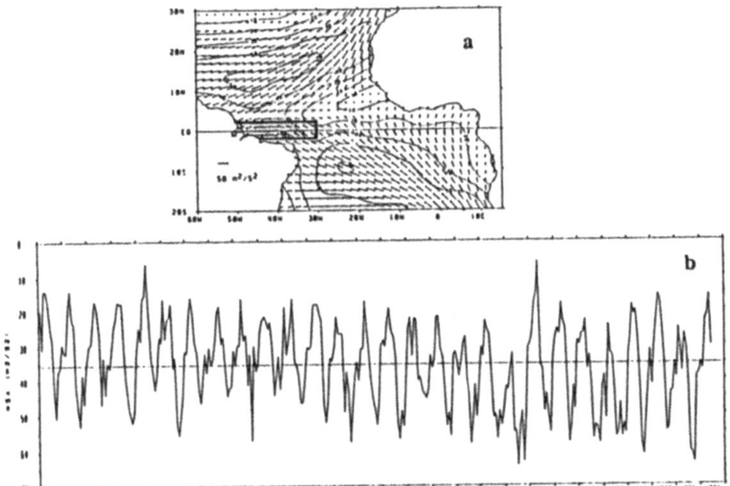

Figure 8 : (a) Total monthly mean of wind stress and (b) time series from January 1964 to June 1991 of monthly zonal wind stress anomalies corresponding to the dotted equatorial zone in the west of the basin. One negative (positive) anomaly indicates an intensification (relaxation) of zonal wind stress in relation to the climatology [updated after Servain & Séva, 1987]

leading North basin by a few months). The Atlantic SST dipole grows progressively during the months surrounding the peak of an ENSO (or anti-ENSO) episode (Servain, 1991).

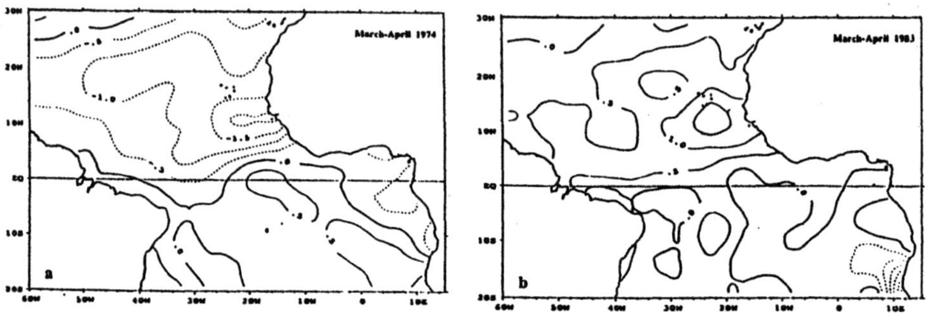

Figure 9 : (a) Observed SST anomaly (in °C) averaged during March-April 1974 and (b) March-April 1983 (b) [after Servain, 1991]

4. RELATIONS WITH NAO

As for SO, the NAO is a large-scale alternation which was first identified in the 1920s by Sir Gilbert Walker. The NAO is a temporal fluctuation of the zonal wind strength across the Atlantic Ocean due to pressure variations in both the subpolar low near Iceland and the subtropical anticyclone belt (Fig 10). Walker (1924) and Walker & Bliss (1932) suggested that a simple index of the NAO is the pressure difference between the Azores and Iceland (Fig 11). Bjerknes (1964) related this oscillation to long-term fluctuations in SST changes over the north and tropical Atlantic sectors. More recently, a series of papers by van Loon & Rogers (1978), Rogers & van Loon (1979), Meehl & van Loon (1979) reviewed this climatic signal for its dominant impact on the weather conditions over Europe during winter.
A possible link between SO and NAO is not clear : Rogers (1984) founds that both oscillations are significantly correlated, while Barnett (1985) suggests the NAO is only a singular expression of internal atmospheric dynamics and is unrelated to the world dominant SLP signal associated with the SO.

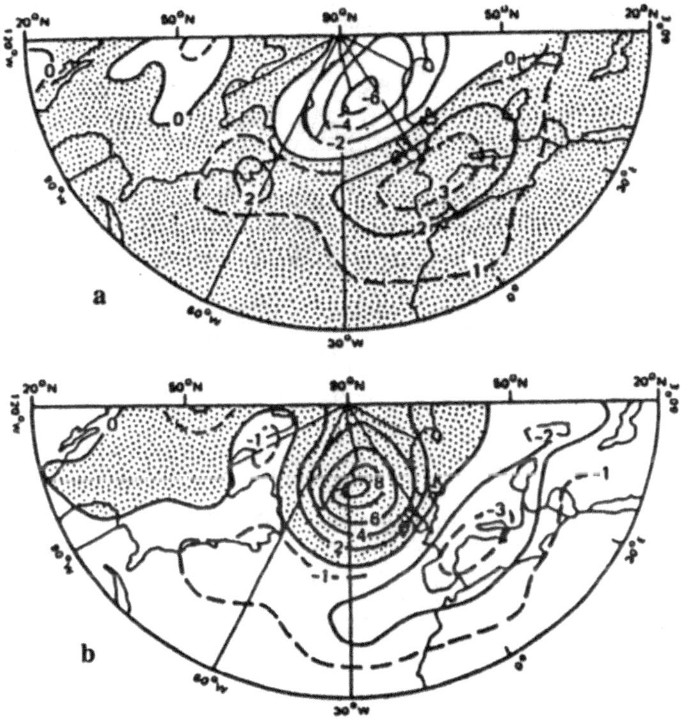

Figure 10 : The extremes of the NAO. Panel a shows the average January SLP anomaly pattern (mb) when the temperature in Greenland is below the mean and that in Norway is above the mean; Panel b is for the inverse case [after van Loon & Rogers, 1978]

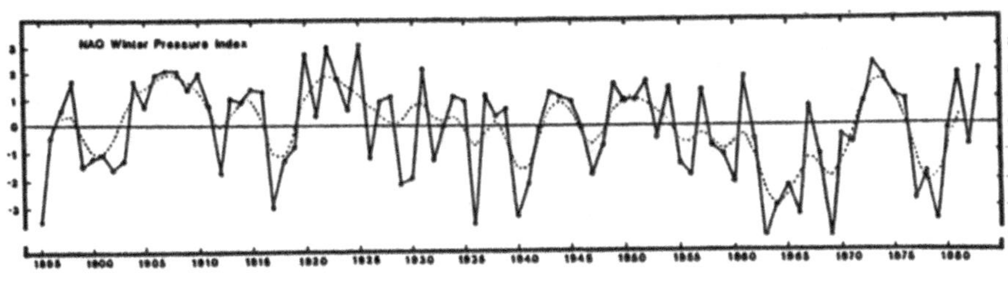

Figure 11 : Winter index of the NAO during years 1895-1993 based on the mean normalized pressure difference between Azores and Iceland [after Rogers, 1984]

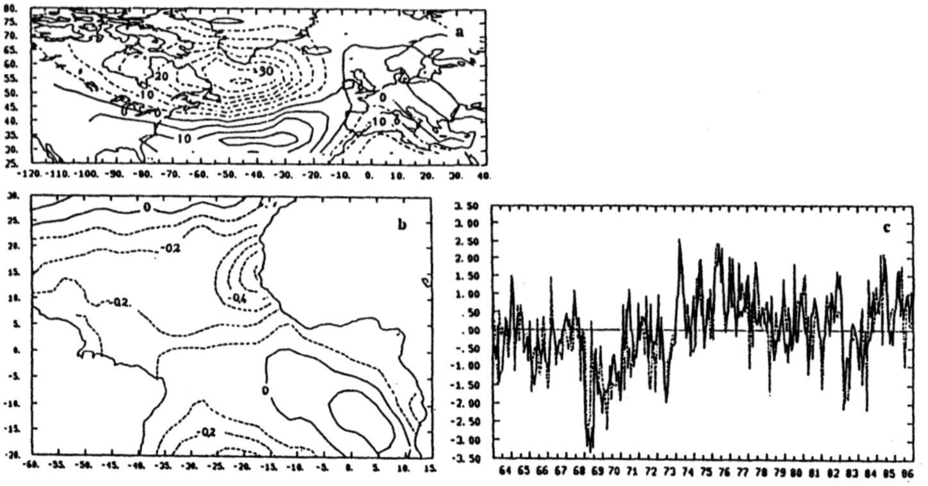

Figure 12 : Canonical correlation analysis at lag 1 month (height leads) between the midlatitude 50 kPa height anomalies and the tropical Atlantic SST anomalies (°C) during 1964-1984: (a) first canonical loadings for height (m) and (b) for SST (°C); (c) canonical variates for SST (solid line) and height (dashed line) (dimensionless unit); the dates correspond to the SST variates, the height variates being shifted 1 month later [after Déqué & Servain, 1989]

Déqué & Servain (1989) found two teleconnection modes between monthly SST anomalies in the tropical Atlantic basin and monthly height anomalies of the 50kPa surface over an area covering the northern midlatitudes from North America to central Europe. One of these linkages, being almost in phase, seems to be controlled by a direct mechanism. An height pattern, fairly similar to that of NAO but rather occurring during the spring-time, is associated with an oceanic thermal response of a single sign in practically the whole tropical basin (Fig 12) : when the westerly circulation is strengthened (weakened), negative (positive) SST anomalies tend to develop in the tropics. It was particularly the case during March-April 1969 (Fig 13) : in March 1969, the midlatitudes westerlies were much lighter than normal (Posey, 1969), and a strong westerly surface-stress anomaly was observed between 20°N and 30°N, which collapsed in April (Picaut et al., 1985). One month later, we

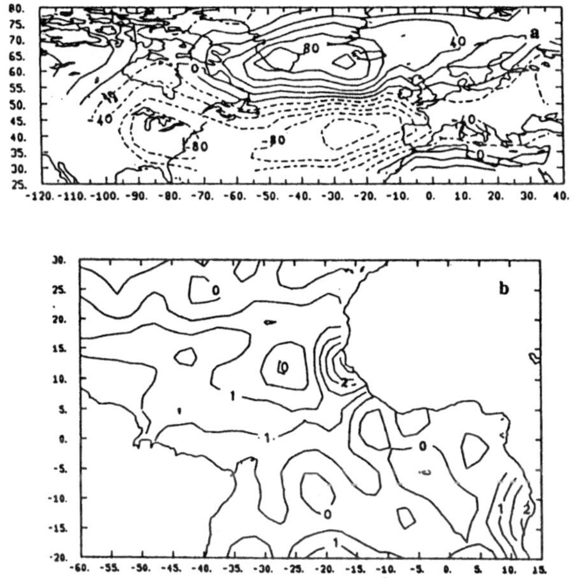

Figure 13 : (a) Observed height anomaly (m) of March 1969 and (b) SST anomaly of April 1969 [after Déqué & Servain, 1989]

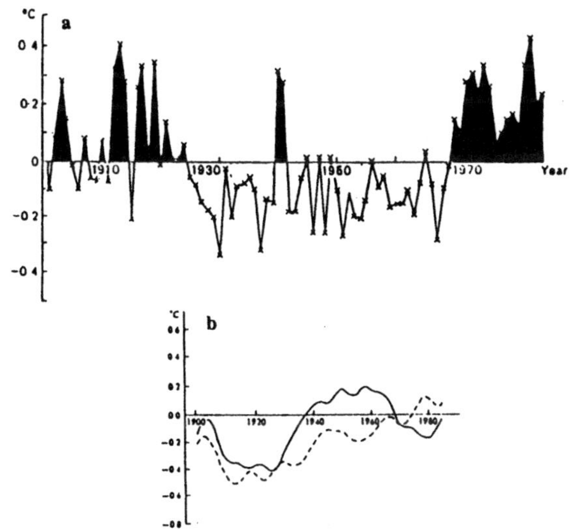

Figure 14 : (a) SST anomaly differences - southern hemisphere minus northern hemisphere. The southern hemisphere included the Indian Ocean and the northern excluded the Indian Ocean. Anomalies were computed relative to the 1951-1980 period, for July to September 1901-1985. (b) SST anomalies (relative to 1951-1980) for the North Atlantic and South Atlantic [after Folland et al., 1986]

observed a general warming in the tropical basin. An important opposite phenomenum of this kind is found in March-April 1974.

5. THE DECADAL VARIABILITY

We already discussed the climatic dipole existing over the tropical Atlantic Ocean. A relation between such interannual mode of variability of SST in the tropical Atlantic and rainfall fluctuations in Brazilian Nordeste and Angola is well established (Moura & Shukla, 1981). Rainfall in the Sahel, however, varies on much longer (decadal) time scales and is correlated with SST changes on a global scale (Figs 14 and 15). This later interhemispheric oscillation that has been recently evidenced by Folland et al. (1986) is especially marked over the Atlantic Ocean. It could result from different ranges in the climatic trends of both hemispheres (Fig 14). For example, the southern Atlantic experienced sustained warming from the beginning of this century, and is currently at its warmest level (Jones, 1988; Servain et al., 1990). Northern Atlantic temperatures increased until 1930-1950, decreases during the 1960s, and increased again during the past 10 years.

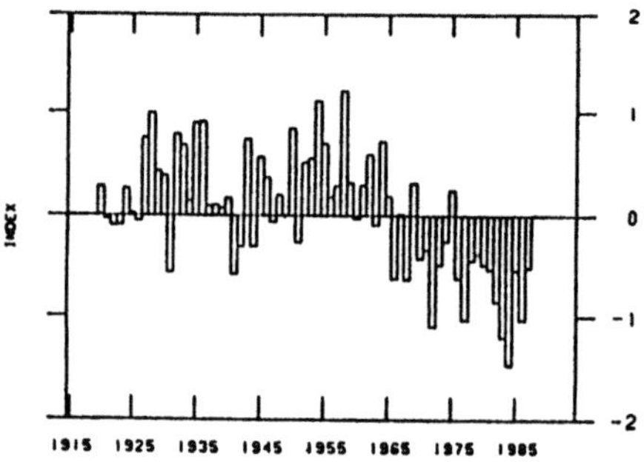

Figure 15 : Time series of averaged standardized precipitation based on data for 13 reporting stations in the western Sahel [updated after Lamb, 1983]

One index illustrating the meridional dipole over the tropical Atlantic was recently proposed by Servain (1991). This index is obtained in averaging standardized monthly SST anomalies over the northern and southern tropical Atlantic basins (the line of separation being 5°N), and taking the difference between the two resulting values (Fig 16). The meridional dipole index clearly exhibits a decadal time scale (Fig 17). Thus, the decadal interhemispheric SST signal that has been evidenced by Folland et al. (1986), and which affects the world Ocean, is particularly marked in the tropical Atlantic sector and is in close relation with the latitudinal position of the ITCZ and rainfall anomaly over the Sahelian region.

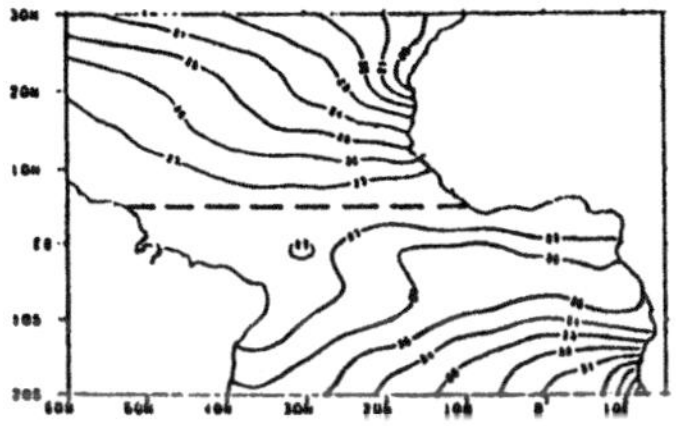

Figure 16 : Total average (1964-1989) of monthly SST (°C) over the tropical Atlantic basin. The broken line at 5N represents approximatively the climatic latitude of the thermal equator [after Servain, 1991]

6. DISCUSSION AND CONCLUSIONS

The interannual climate variability over the tropical Atlantic Ocean was until recently commonly considered as weak, compared in particular to the seasonal climate variability, and mostly under the dependance of worldwide signals like ENSO. However it was known that the interannual signal of the zonal wind stress prevails in the western equatorial zone, which is of outstanding importance for inducing any eastward remote dynamical responses along the equatorial guideline (Moore et al., 1978; Servain et al., 1982). Attempts have been made to relate the ENSO signal to the climatic variability in the Atlantic sector (Fig 5). The Atlantic Ocean is directly influenced by the SO through the global perturbation of the tropical low atmosphere system. In 1983, for example, the unusual El Niño conditions over the tropical Pacific Ocean contributed to intense trade winds and relatively low SST over the tropical Atlantic Ocean. In general, El Niño conditions in the Pacific have a significant effect on the Amazon basin where rainfall is reduced. But, in addition to being directly influenced by the SO, the tropical Atlantic has a delayed interannual mode of oscillation that is similar to but distinct from the SO. The atmospheric convergence onto the region of unusually high SST in the eastern Pacific during the mature phase of an El Niño causes intense trade winds blowing over the western Atlantic, and corresponds to a build up phase of an "El Niño-like" event in the Atlantic. When the trade winds return to normal in the Pacific it corresponds to a wind relaxation in the Atlantic. Then a warm event could develop in that region with a similar physics than in the Pacific. It is what happened in 1984, one year after the huge 1982-1983 Pacific El Niño.

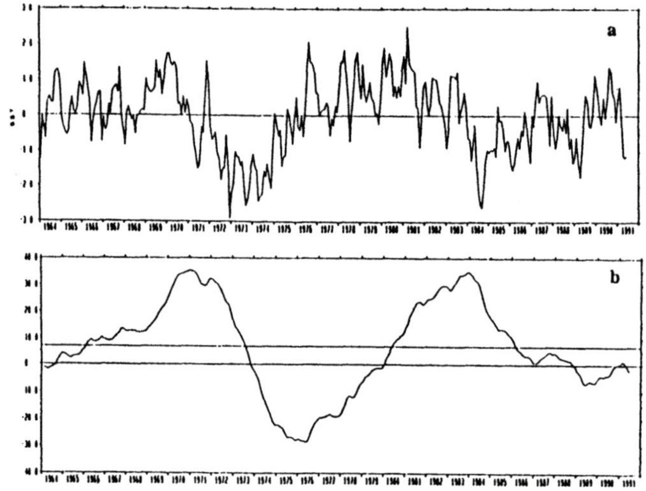

Figure 17 : Monthly SST dipole index from January 1964 to June 1991 computed taking the difference between monthly SST averaged between 28N and 5N and monthly SST averaged between 5N and 20S; (a) is the time series of the individual standardized values of the dipole, (b) is the time series of the cumulated (from January 1964) standardized values of the dipole [updated and added after Servain, 1991]

Another caracteristic of these warm events connected to ENSO, is their relation with the latitudinal position of the ITCZ. In both the Pacific and the Atlantic a warm event (El Niño or "El Niño-like") is associated with an abnormal southward position of the ITCZ.

Bjerknes (1964) documented the meteorological changes associated with the decadal SST fluctuations over the northern Atlantic of opposite sign with the southern Atlantic (dipole). He hypothezised that these long term changes could be in relation with changes of heat fluxes that flow from South to North all over the meridionally extented Atlantic Ocean. Bjerknes also described extensively variations in sea level pressure, winds and SST that affect only the North Atlantic on relatively short intra-annual time scale and which have again a dipole structure, i.e. the NAO. The NAO is caracterized by a correlation betwen an unusually deep Icelandic low pressure and a strong North Atlantic (Azores) anticyclone. Because the NAO seems to be independant of ENSO, but is also related to the tropical Atlantic variability, the complexity of the climate study in this region increases.

Thus the tropical Atlantic Ocean has a rich spectrum of climate variability that covers time scales from months to decades, at least. It should be noted that practically the entire spectrum of variability involves anomalous north-south movements of ITCZ. Such movements are also a main feature of ENSO but have received far less attention than the east-west movements of the convergence zone over the western tropical Pacific.

A central question is therefore what processes cause the seasonal, interannual and decadal movements of the ITCZ ? It seems, at the first order, that these important latitudinal displacements of the ITCZ in the Atlantic Ocean are related to the meridional extension of the basin, which is largely open to the high latitudes both in the north and the south hemispheres, but is zonally limited in the equatorial region. Surface waters become increasingly saline as they flow from the southern Ocean. These inflowing waters are compensated-for by the outflow of relatively fresh deep water. Such process has been dubbed the "Atlantic conveyor belt". At the sea surface in the equatorial region, such interannual signal linked to this mode of variability could possibly be documented by the observations of the abnormal meridional displacements of the ITCZ associated with aforementioned climatic features like the meridional dipole.

Thus, the Atlantic Ocean seems to be a major player in the establishing global climate change at various time scales. However the present understanding of the Atlantic's role in climatic change is much less developed than for the Pacific. The aim of future climate programs, as for instance the Atlantic Climate Change

Program (ACCP) (NOAA-ACCP Report, 1990), WOCE, TOGA and post WOCE and TOGA programs would fill this gap.

REFERENCES

Barnett TP (1985) Variations in near-global sea level pressure. J Atmos Sc 42: 478-501

Bjerknes J (1964) Atlantic air-sea interaction. Adv Geophy 10: 1-82

Busalacchi AJ, Picaut J (1983) Seasonal variability from a model of the tropical Atlantic Ocean. J Phys Oceanogr 13: 1564-1588

Daniels JM (1985) Spatial and temporal variations of large scale tropical circulations from satellite outgoing longwave radiation. Thesis of Master of Science, University of Maryland, USA: 82 pp

Déqué M, Servain J (1989) Teleconnections between tropical Atlantic sea surface yemperatures and midlatitude 50 kPa heights during 1964-1986. J Climate 2: 929-944

Folland CK, Parker DE, Ward MN, Colman AW (1986) Sahel rainfall, northern hemisphere circulation anomalies and worldwide sea temperature changes. In: Long range forecasting and climate research memorandum No. 7a. Meteorological Office (Met O 13), London Road, Bracknell, Berkshire RG12 2SZ: 51 pp

Hastenrath S, Heller L (1977) Dynamics of climatic hazards in northeast Brazil. Q J R Meteorol Soc 103: 77-92

Hastenrath S, Kaczmarczyk EB (1981) On spectra and coherence of tropical climate anomalies. Tellus 33: 453-462

Horel JD, Kousky VE, Kagano MT (1986) Atmospheric conditions in the Atlantic sector during 1983 and 1984. Nature 322: 248-251

Jones PD (1988) Hemispheric surface air temperature variations: recent trends and an update to 1987. J Climate 6: 654-660

Katz EJ, Hisard P, Verstraete JM, Garzoli S (1986) Annual change of sea surface slope along the equator of the Atlantic Ocean in 1983 and 1984. Nature 322: 245-247

Kidson JW (1975) Tropical eigenvector analysis and the Southern Oscillation. Mon Wea Rev 103: 187-196

Lamb PJ (1978a) Large-scale tropical Atlantic surface circulation patterns associated with sub-Saharan weather anomalies. Tellus 30: 240-251

Lamb PJ (1978b) Case studies of tropical Atlantic surface circulation patterns during recent sub-Saharan weather anomalies: 1967 and 1968. Mon Wea Rev 106: 482-491

Lamb J (1983) Rainfall in Subsaharan West Africa during 1941-1983. Z Gletscherk Glazialgeol 21: 131-139

Meehl GA, van Loon H (1979) The see-saw in winter telperatures between Greenland and Northern Europe. Part III: Teleconnections with lower latitudes. Mon Wea Rev 107: 1095-1106

Merle J, Fieux M, Hisard P (1980) Annual signal and interannual anomalies of sea surface temperature in the eastern equatorial Atlantic. Gate Sup. II, Deep-Sea Res 26: 77-101

Moore DW, Hisard P, McCreary JP, Merle J, O'Brien JJ, Picaut J, Verstraete JM, Wunsch C (1978) Equatorial adjustement in the eastern Atlantic. Geophys Res Lett 5: 637-640

Moura AD, Shukla J (1981) On the dynamics of droughts in northeast Brazil: Observations, theory and numerical experiments with a general circulation model. J Atmos Sc 38: 2653-2675

NOAA-ACCP Report (1990) Atlantic climate change program (ACCP) science plan. Special Report No. 2 prepared by the Atlantic Climate Change Science Working Group: 29 pp

Philander SCH (1979) Variability of the tropical oceans. Dyn Atmos Oceans 3: 191-208

Picaut J, Servain J, Lecomte P, Séva M, Lukas S, Rougier G, (1985) Climatic atlas of the tropical Atlantic wind stress and sea surface temperature 1964-1979, Unviversité de Bretagne Occidentale, Brest, France [Available from Laboratoire d'Océanographie Physique, Faculté des Sciences,29200 Brest, France]: 467 pp

Posey JW (1969) The weather circulation of March 1969: A very cold month with a developing flood threat in the upper Midwest. Mon Wea Rev 97: 464-470

Quiroz RS (1983) The climate of the "El Niño" winter of 1982-83, a seasonal of extraordinary climatic anomalies. Mon Wea Rev 111: 1685-1706

Rogers JC (1984) The association between the North Atlantic Oscillation and the Southern Oscillation in the northern hemisphere. Mon Wea Rev 112: 1999-2015

Rogers JC, van Loon H (1979) The see-saw in winter temperatures between Greenland and Northern Europe. Part II: Some oceanic and atmospheric effects in middle and high latitudes. Mon Wea Rev 107: 509-519

Servain J (1984) Réponse océanique à des actions éloignées du vent dans le Golfe de Guinée en 1967-1968. Oceanol Acta 7: 297-307

Servain J (1985) Relation entre la position latitudinale de la zone de convergence intertropicale et la température de surface de la mer dans l'Atlantique tropical. Veille Climat Satell 9: 34-36

Servain J (1991) Simple climatic indices for the tropical Atlantic Ocean and some applications. In press in J Geophys Res

Servain J, Legler DM (1986) Empirical orthogonal function analyses of tropical Atlantic sea surface temperature and wind stress: 1964-1979. J Geophys Res 91: 14181-14191

Servain J, Séva M (1987) On relationships between tropical Atlantic sea surface temperature, wind stress and regional precipitation indices: 1964-1984. Ocean-Air Interactions J 1: 183-190

Servain J, Picaut J, Merle J (1982) Evidence of remote forcing in the equatorial Atlantic Ocean. J Phys Oceanog 2: 457-463

Servain J, Picaut J, Busalacchi AJ (1985) Interannual and seasonal variability of the tropical Atlantic Ocean depicted by sixteen years of sea-surface temperature and wind stress. In: Coupled Ocean-Atmosphere Models, Nihoul JCJ (ed): 211-237

Servain J, Séva M, Rual P (1990) Climatological comparison and long-term variations of sea surface temperature over the tropical Atlantic Ocean. J Geophy Res 95: 9421-9431

Stommel H (1979) Ocean warming of western Europ. In: Proc Natl Acad Sci USA 76: 2518-2521

Trenberth KD, Shea DJ (1987) On the evolution of the Southern Oscillation. Mon Wea Rev 115: 3078-3096

van Loon H, Rogers JC (1978) The see-saw in winter temperatures between Greenland and Northern Europe. Part. I: General description. Mon Wea Rev 106: 296-310

Walker GT (1924) Correlations in seasonal variations of weather, IX. in: Mem Indian Meteor Dept 24: 275-332

Walker GT, Bliss EW (1932) World weather V. In: Mem Rot Meteor Soc 44: 53-84

LOW-FREQUENCY VARIABILITY AND BLOCKING AS DIAGNOSTIC TOOLS FOR GLOBAL CLIMATE MODELS

Stefano Tibaldi
Università di Bologna
Dipartimento di Fisica
Gruppo Dinamica Atmosferica
Via Irnerio 46, 40126 Bologna, Italy

Atmospheric General Circulation Models (AGCMs) are being employed more and more extensively for climate impact studies, often coupled with Ocean General Circulation Models (OGCMs) to give what is commonly referred to as a Global Climate Model.

Such Global Climate Models are often used to predict the influence of man-produced greenhouse gases (CO_2, CFCs, CH_4, and so on) upon the earth's climate on a time scale that ranges from a few decades to a few centuries. Several "scenarios" are usually explored, both with the so-called "equilibrium" integrations, in which the increased concentration of greenhouse gases is present since the beginning of the experiment, and with "time-dependent" integrations, in which the desired level of optically active gases is reached in a time representative of the current and future emission trends. Many of such experiments have provided the well-known global temperatore increase estimates ranging from 0 to 5 degrees centigrades.

It is not the purpose of this brief contribution to discuss at length and in its more profound implications the quantitative uncertainty attached to such global warming estimates. Rather, the starting point to show how atmospheric low-frequency variability can constitute the basis for devising very powerful

NATO ASI Series, Vol. I 6
Prediction of Interannual Climate Variations
Edited by J. Shukla

model diagnostic tools will be the need to "specialize" such global temperature "predictions"in terms of regional diversification. Such need is felt almost universally among the perspective users of such climatological model "forecasts", and has prompted further work in the field of the so-called "Regionalization of global Climate Predictions". Such work has proceeded so far mainly along two parallel approaches: the use of nested limited-area models (LAMs) and the use of statistical techniques, essentially based on the MOS technique used in meteorology to derive local forecasts of "difficult" meteorological parameters, like for example, precipitation.

Recent work (Tibaldi and Molteni, 1990; Tibaldi et al, 1991a; Tibaldi et al, 1991b) has shown that atmospheric blocking is often very poorly represented in General Circulation Models (GCMs) used for medium- and extended-range predictions, to the extent that the majority of blocking onsets is missed by state-of-the-art global forecasting models if it takes place more than 3 to 4 days into the forecast. The geographical (space) and seasonal (time) variability of blocking anticyclones occurrence, however, participates in determining much of the space-time variability of mid-latitude and subtropical precipitation patterns and especially the largest anomalies responsible for relatively short-lived drought periods.

If GCMs are to be employed with any degree of success as climate forecasting tools, their space-time representation of blocking patterns typical of present-day atmospheric climate must be reasonable, otherwise little trust can be put on any increased-CO_2 scenario climate prediction. Any further regional climate prediction, in itself a problem fraught with difficulties (e.g. Grotch and MacCracken, 1991 and Gutowski et al, 1991), would be affected by even greater uncertainties, becoming in fact devoid of any meaning.

As a preliminary attempt to diagnose the ability of Climate Models to correctly represent the space-time variability of atmospheric blocking, two twenty-year annual cycle integrations of the ECHAM T21 model (Roeckner et al. 1989) have been analyzed. The same blocking index as in Tibaldi and Molteni (1990) has been computed (that is a modification of the original Lejenas and Okland index, Lejenas and Okland, 1983), with the only difference that the index has been computed on 5-day time mean 500 hPa fields, instead of daily 12GMT fields. Northern Hemisphere winter (December to February) only will be briefly discussed here (work is in progress for the other seasons of the year). To provide statistically solid observational data for verification purposes, a long time series of NMC 500 hPa 5-day mean fields (1954 to 1980) has been extended to 1990 by making use of ECMWF operational analyses, to provide a 37 winter dataset on which the observed index was computed.

Two model integrations have been analyzed; they both use climatological sea surface temperatures as lower boundary conditions and were originally produced as "zero level" control runs. The first (ECHAM-1) was produced using both envelope orography (Wallace et al. 1983) and Gravity Wave Drag (GWD, e.g., Palmer et al. 1986) and the second (ECHAM2-) with mean orography and no GWD. Figure 1 shows the blocking frequency distribution as a function of longitude. Panel a) shows observations, while panels b) and c) the results coming from the two model integrations. In panel a) it is easy to recognize the two NH quadrants in which blocking is more likely to occurr: Euro-Atlantic Blocking and Pacific Blocking.

Regarding model results, two main considerations should be maid: the first is that the overall results are not as bad as one could have expected on the basis of past experience with comparatively lower resolution integrations. There is in fact ample evidence that to achieve good short-, medium- and extended-range forecasts of blocking onset and maintenance model resolution is of

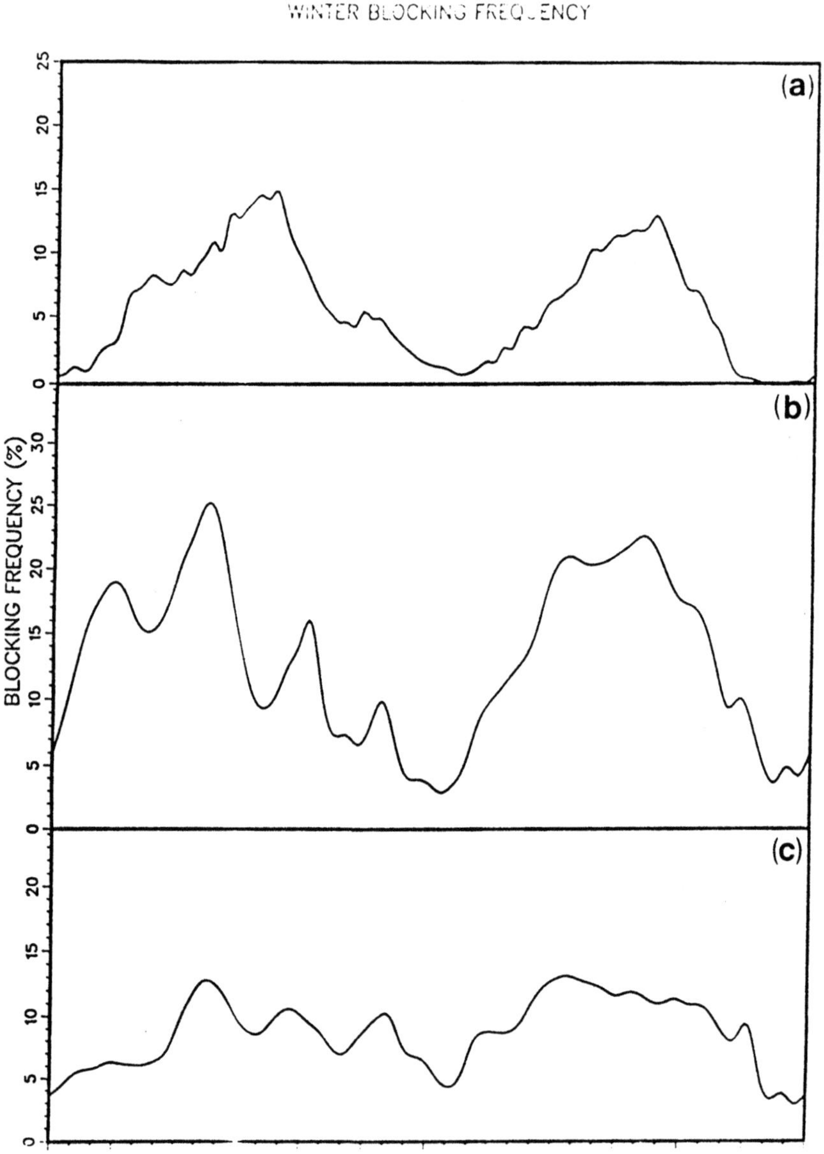

Figure 1: Frequency of Northern Hemisphere winter (DJF) blocking as a function of longitude as measured by the Lejenas and Okland (1983) index modified by Tibaldi and Molteni (1990): a) Observed (1954-1990); b)ECHAM-1 20-year integration, envelope orography and GWD included; c)ECHAM-2 20-year integration, mean orography, no GWD.

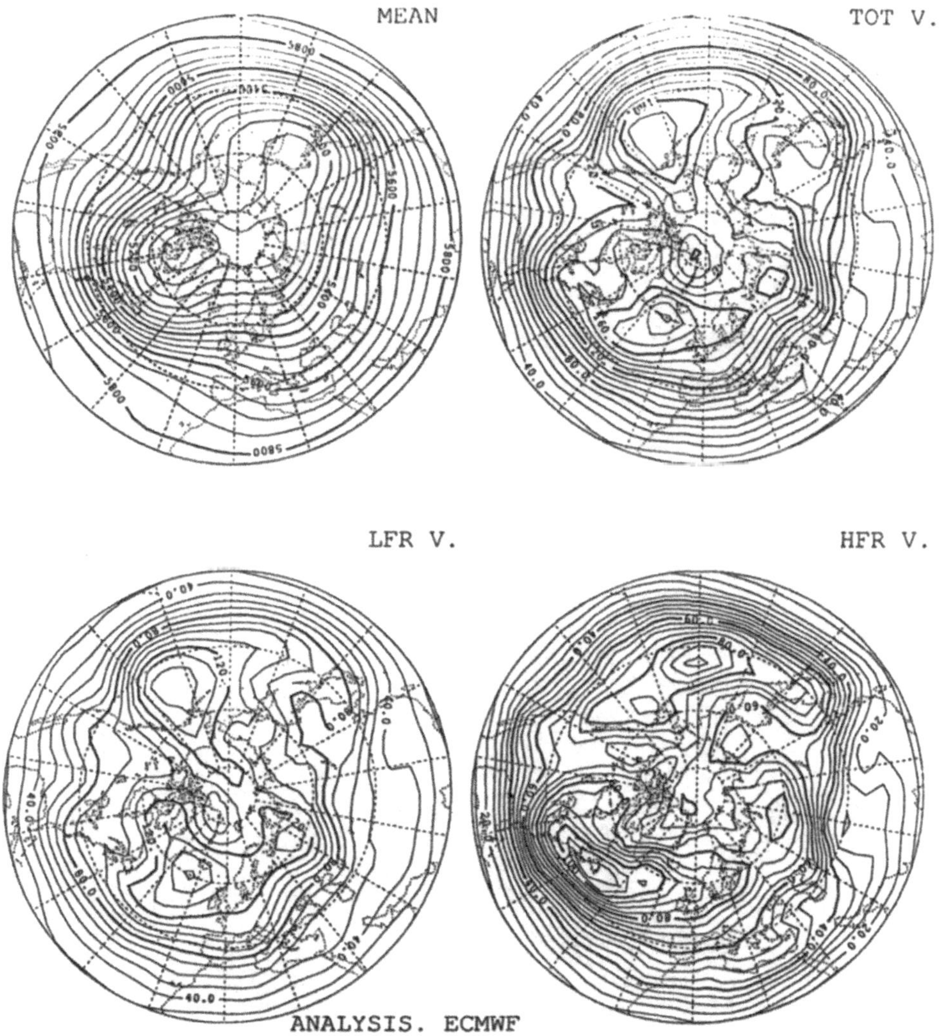

Figure 2: Observed winter (DJF) Northern Hemisphere 500 hPa geopotential height fields, ECMWF operational analyses, 1980-1990; a) mean, b) total variance, c) low-frequency (periods longer than 5 days) variance, d) high-frequency (periods shorter than 5 days) variance.

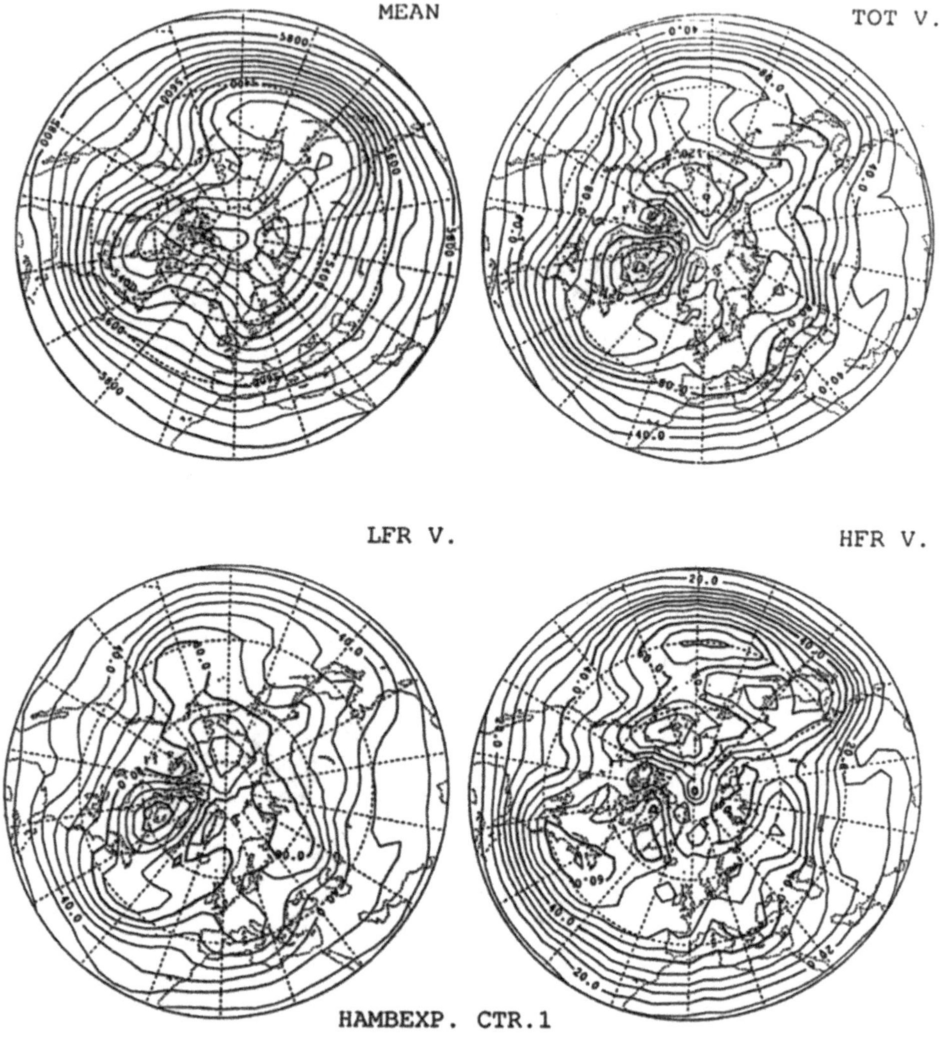

Figure 3: As Figure 2, but for ECHAM-1 20-year integration.

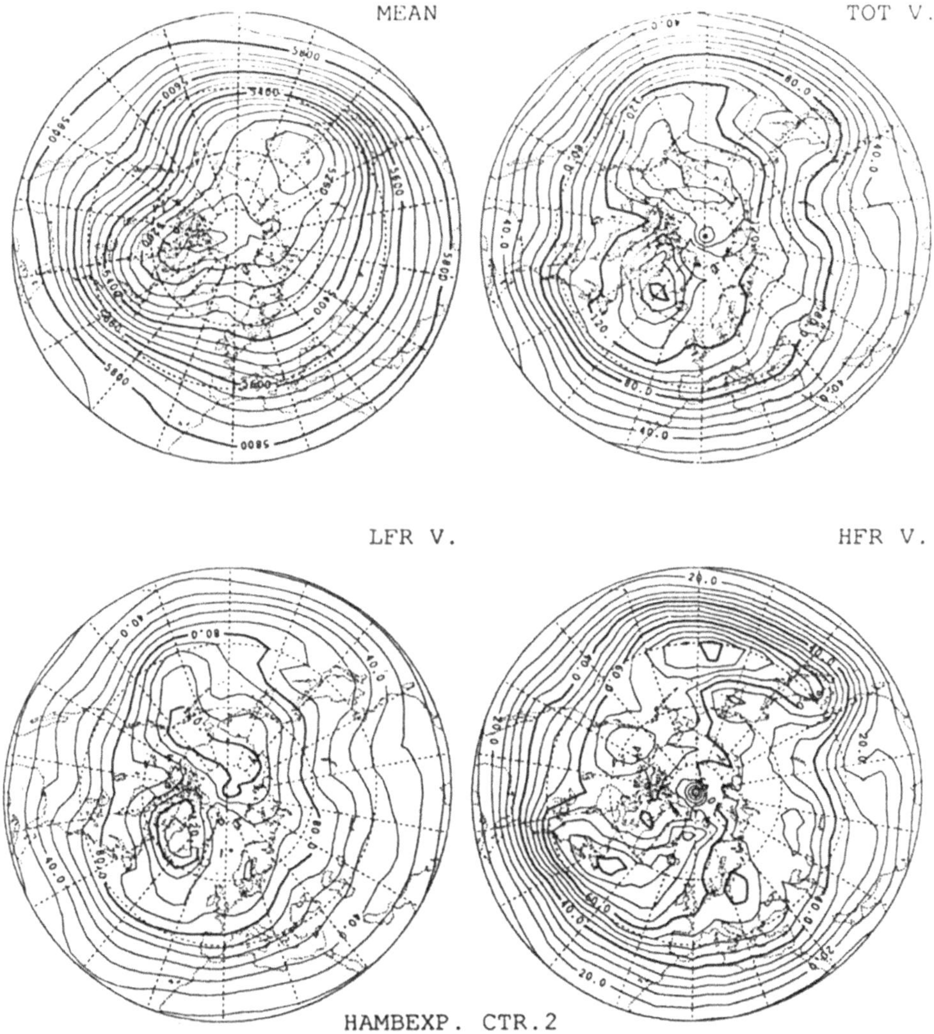

Figure 4: As Figure 2, but for ECHAM-2 20-year integration.

importance (e.g. Bengtsson, 1981, Tibaldi and Ji, 1982, Miyakoda and Sirutis, 1990). This could possibly be due to a better model representation of the effects of synoptic eddies, often considered responsible for blocking maintenance (e.g. Green, 1977; Shutts, 1883).

The second is that ECHAM-2 performance is much better than ECHAM-1 in terms of overall NH winter blocking frequency (ECHAM-1 being overestimated), but ECHAM-1 succeeds in having a better separation between the two main geographical frequency maxima (ECHAM-2 having difficulties in producing two clearly isolated frequency maxima corresponding to the observed Euro-Atlantic and Pacific peaks).

Since blocking is one of the main sources of low frequency variability in the NH winter and the problems GCMs (and, in particular, low-resolution ones) have in representing both low- and high-frequency variability are well known, it is interesting to observe how the models' behaviour in representing blocking is well reflected in their simulations of geopotential height variance; see Figures 2, 3 and 4 for obsevations, ECHAM-1 and ECHAM-2 respectively (here observations are based on 10-year ECMWF operational daily analyses). Again ECHAM-2 performs much better, in agreement with envelope orography and GWD known mechanisms of action and their effects in low-resolution GCMs (Tibaldi, 1986; Miller et al, 1989). In particular, ECHAM-2 has a much higher level of both low- and high-frequency variance (although still too low compared to observations) and an overall more realistic mean 500 hPa geopoptential height field.

It would appear, therefore, from these preliminary results (however limited in scope and applicability) that low-resolution GCMs have difficulties in representing satisfactorily **both** the overall NH winter blocking frequency and its correct geographical variability and that the representation of orography could play a role in explaining such difficulties. Since it is well known that compensation of model errors is taking place to a considerable degree in low-resolution GCMs and that this is related to representation of orographic effects (Tibaldi, 1986; Miller et al, 1989), it is reasonable to assume that modelling of blocking could be affected shortcomings of similar nature.

Before downscaling of Global Models' otput or Regional Climate Modelling are massively performed in an organized way, it would appear therefore wise to extend such diagnostic work on blocking in AGCMs to all seasons of both hemispheres, to more than one AGCM and, most important of all, to higher resolution atmospheric climate runs with the best possible boundary forcing (i.e. runs with observed sea surface temperatures). Some lack of consistency between results from statistical studies and from single case studies and between GCMs used for weather forecasting and GCMs used for Climate Prediction should also be resolved by investigating the sensitivity of results to the exact specification of the blocking index used, and at the same time searching for better objective blocking indices.

REFERENCES

Bengtsson, L., 1981: Numerical prediction of atmospheric blocking. A case studied. Tellus, 33A, 19-42.

Green, J.S.A., 1977: The wether during July 1976: some dynamical considerations on the drought. Weather, 32, 120-126.

Grotch, S.L. and M.C. MacCracken, 1991: The Use of General Circulation Models to Predict Regional Climatic Change, J. Climate, 4, 286-303.

Gutowski, W.J., D.S. Gutzler and W.-C. Wang, 1991: Surface Energy Balances of Three General Circulation Models: Implications for Simulating Regional Climate Change, J. Climate, 4, 121-133.

Lejenas, H. and H. Okland, 1983: Characteristics of Northern Hemisphere blocking as determined from a long time series of observational data. Tellus, 35A, 350-362.

Miller, M., T.N. Palmer and R. Swinbank, 1989: Parametrization and influence of subgridscale orography in general circulation and numerical weather prediction models. Meteorol. Atmos. Phys., 40, 84-109.

Miyakoda, K. and J. Sirutis, 1990: Subgrid scale physics in 1-month forecasts. Part II: Systematic error and blocking forecasts. Mon. Wea. Rev., 118, 1065-1081.

Roeckner, E., L. Duemenil, E. Kirk, F. Lunkeit, M. Ponater, B. Rockel, R. Sausen and U. Schlese, 1989: The Hamburg version of the ECMWF model (ECHAM). GARP rep. No. 13, WMO, Geneva, WMO/TP No. 332.

Palmer, T.N., Shutts, G.J. and Swinbank, R., 1986: Alleviation of a systematic westerly bias in general circulation and numerical weather prediction models through an orography gravity wave drag parametrization. Q. J. R. Meteorol. Soc., 112, 1001-1039.

Shutts, G.J., 1983: The propagation of eddies in diffluent jet streams: eddy vorticity forcing on 'blocking' flow fields. Q. J. R. Meteorolol. Soc., 109, 737-761.

Tibaldi, S., 1986: Envelope orography and maintenance of the quasi-stationary circulation in the ECMWF global models. Advances in Geophysics, 29, 339-374.

Tibaldi, S. and L.R. Ji,1982: On the effect of model resolution on numerical simulation of blocking. Tellus, 35A, 28-38.

Tibaldi, S. and F. Molteni, 1990: On the operational predictability of blocking. Tellus, 42A, 343-365.

Tibaldi, S., A. Navarra and F. Molteni, 1991a: Seasonal variations of blocking frequency and its medium-range predictability. Submitted for publication.

Tibaldi, S., E. Tosi, A. Navarra and F. Molteni, 1991b: Southern Hemisphere blocking and its predictability. Submitted for publication.

Wallace, J.M., S. Tibaldi and A. Simmons, 1983: Reduction of systematic forecast errors in the ECMWF model through the introduction of an envelope orography. Q. J. R. Meteorolo. Soc., 109, 683-717.

OPERATIONAL SEASONAL PREDICTION AT NMC

P. A. Arkin
National Meteorological Center
NWS/NOAA
5200 Auth Road
Camp Springs, MD 20746
USA

1. INTRODUCTION

The National Meteorological Center (NMC) of the U. S. National Weather Service (NWS) is responsible for the production of operational extended range predictions of general climatic conditions over the U. S. in several time ranges. These differ from the forecasts issued by the NWS for periods out to 5 days in that they predict time averaged departures from normal of surface temperature and precipitation, rather than specific weather events. Such predictions are clearly near the frontiers of the state of the art, and methods of improving them are one of the topics of the present workshop. In this paper, we will discuss the current operational practices used at NMC to produce seasonal forecasts. We will begin with a brief description of the background and historical context, will then present the current process, and will conclude with some cautious remarks regarding the future of such operational predictions.

NATO ASI Series, Vol. I 6
Prediction of Interannual Climate Variations
Edited by J. Shukla

2. BACKGROUND/HISTORY

Certain physical phenomena are easily predictable very far into the future. The phases of the moon, the times of the solstices and equinoxes and other features that depend upon planetary orbits are so near to being strictly periodic, on the time scales of interest to people, that knowledge of their past history is sufficient for usable predictions of their future. Physical systems that behave in this manner are referred to as deterministic, and are characterized by damping of small perturbations. They are clearly predictable in a practical sense.

Physical systems in which small perturbations tend to amplify to the scale of the principal motions in the system over time scales similar to those for which predictions are desired are sometimes referred to as chaotic. They are non-periodic and the details of future motions cannot in general be usefully predicted from a knowledge of the state of the system at any particular time. The atmosphere is such a system, and its winds and temperatures cannot be predicted in detail for more than a few weeks. In fact, the current practical limit of useful predictability is on the order of one week. While the full climate system consists of far more than just the atmosphere, it seems highly likely that it too is chaotic with limited deterministic predictability.

However, many important uses exist for predictions of longer than a week or two. Many of these uses could be satisfied with significantly less information than the users of typical 1-2 day weather predictions require. For example, even modestly skillful predictions of seasonal mean temperature are extremely helpful to utility companies in the planning of resource allocation. Fortunately, there exists considerable evidence (discussed in several of the other papers in this volume) that such predictions are possible, largely due to the interactions of components with varying time scales. For example, some variations in tropical sea surface temperatures (SSTs) are observed to be associated with characteristic anomalous circulation patterns.

This mix of requirements and practicality has led to the current pragmatic solution, in which forecasts for periods farther in the future than 5 days are issued, but in a different form than those for the shorter term. The NWS produces forecasts for the period covering days 6-10 from the starting time three times each week, for a one month period twice each month, and for a three month season once each month. These predictions are probabilistic and categorical in format. The methods used are different for all three.

The history of extended range prediction at NMC (this discussion relies heavily on Wagner, 1989) began during the 1940's. At that time, 5-day forecasts (initially for days 3-7) began to be made under the leadership of Jerome Namias. Monthly forecasts began at about the same time. The period for 5-day forecasts has evolved to the current days 6-10 as the skill of objective numerical weather forecasts has advanced, and the role of numerical models in the production of that forecast has increased over time. The monthly forecasts are made for calendar months and for the mid-month to mid-month periods as well, with a lead time (the length of time between the making of the forecast and the beginning of the period for which it is valid) of several days. The principal change here over the years has been a modest decrease in the lead time.

The seasonal forecasts were first issued to the public, following a long testing period (Namias, 1964), in 1974. The initial forecast was for Northern Hemisphere summer (June-August, or JJA), and for the first 6 years, forecasts were issued only for the meteorological seasons (JJA, September-November or SON, December-February or DJF and March-May or MAM). Beginning with the October-December 1980 season, these forecasts were issued monthly. In this paper, we will restrict our discussion of methods to the seasonal forecast process.

3. SEASONAL FORECAST PROCESS

For shorter time scales, a physically based numerical model of the atmosphere constitutes the principal component of the guidance used to produce forecasts. While the model output is relatively less critical in the production of the monthly forecasts, it still plays a significant role. However, in the seasonal forecast process, such forecasts contribute little. Experience has shown that a wide variety of objective tools can be helpful in producing seasonal forecasts, but that the subjective interaction of several forecasters with these tools and with each other is critical.

The current operational seasonal forecasts are developed in several stages. In the first stage, three to five forecasters individually and independently generate forecast maps. This process begins with each forecaster using a variety of objective indicators to deduce the likely location of anomaly extrema in the prognostic 700 mb height field. The use of 700 mb, rather than some higher level, is due to its longer record of high quality analyses in the Northern Hemisphere. The objective tools available to each forecaster include lag teleconnection and significance maps for leads of 1 month, 1-4 seasons, and selected longer leads (see Wagner, 1989, for more details), the current state of tropical Pacific SSTs and their documented impacts on North American climate (Ropelewski and Halpert, 1986, 1987, 1989), and other factors such as regional variations in persistence. Each forecaster uses this information subjectively to derive a set of, in his opinion, most probable height anomalies.

Each forecaster then identifies a set of past seasons that he feels best accomodate the projected extrema and the clues from which they were derived. This process is generally subjective, although there simple objective tools are available to help in the matching of specific features with past seasons. The number of historical analogues chosen varies among forecasters, and individual forecasters may use different numbers in different seasons. The observed anomalies during the selected analogue seasons are then used to derive categorical (above, normal or below) forecasts of temperature (T) and precipitation (P). This derivation is also a subjective process, consisting basically of compositing of the observed T and P anomalies with subjectively assigned variable weights. In practice, of course, the forecaster tries to identify anomalous T and P features in the historical data which relate, in his opinion, particularly directly to the circulation features he/she expects during the season. His/her categorical forecast will emphasize those features, and will generally be conservative elsewhere.

The forecaster then uses a set of objectively derived T and P predictions to modify his/her categorical forecasts. Many of the indicators used subjectively as described above are also used in an objective statistical prediction scheme developed by

Livezey and Barnston (1988) for DJF and extended to other seasons by Barnston and Livezey (1989). This analog/antilog prediction scheme yields categorical T forecasts, together with indications of expected skill. Certain of the other tools described earlier also can be used to derive objective, semi-quantitative predictions of T or P in selected locations or seasons, as well as information regarding recent biases in the available products (van den Dool and Nap, 1985; van den Dool et al., 1986; van den Dool, 1988). All of these objective products are used subjectively by the forecaster to modify his/her categorical T and P forecasts (Livezey et al., 1990).

Once the 3-5 individual forecasters have completed this process, an open discussion among them and other interested parties is held. The discussion is coordinated by the forecaster responsible for the prediction for the season in question (the lead forecaster). After each forecaster has presented and explained his/her predictions, a final modification of each individual forecast is made by the forecaster. Following this, the individual forecast maps are digitized and the three categories are assigned numerical values (below - 1, normal - 2, above - 3). The categorical predictions are then objectively composited through simple averaging of the categories at a set of station locations.

The consensus T forecast produced by this process is the final categorical version. However, a final modification is applied to the consensus P forecast. The consensus T prediction is used objectively to select 5 analogs from the period 1890-present. The observed P during those 5 seasons is used subjectively by the lead forecaster to modify the consensus P prediction. These final categorical forecasts of seasonal T and P anomalies are then converted to probabilities. This conversion is a subjective process, performed by the lead forecaster, which considers the various inputs, including the projected skills and errors when available, and takes into account the historical record of success in different areas and seasons (Livezey, 1990). An example of the final forecasts for JJA 1990 is shown in fig. 1.

The skill of the operational forecasts is difficult to measure. Not only does it vary with season, location, flow pattern and the state of the El Nino/Southern Oscillation (ENSO) phenomenom, but it is hard to obtain a statistically adequate number of samples because the forecasts are made relatively infrequently. The skill of the operational seasonal predictions of T and P over the entire period for which they have been produced (including the experimental portion) is shown in fig. 2. Also shown for contrast is the skill of the operational monthly T and P forecasts for the period 1981-1988. It is clear

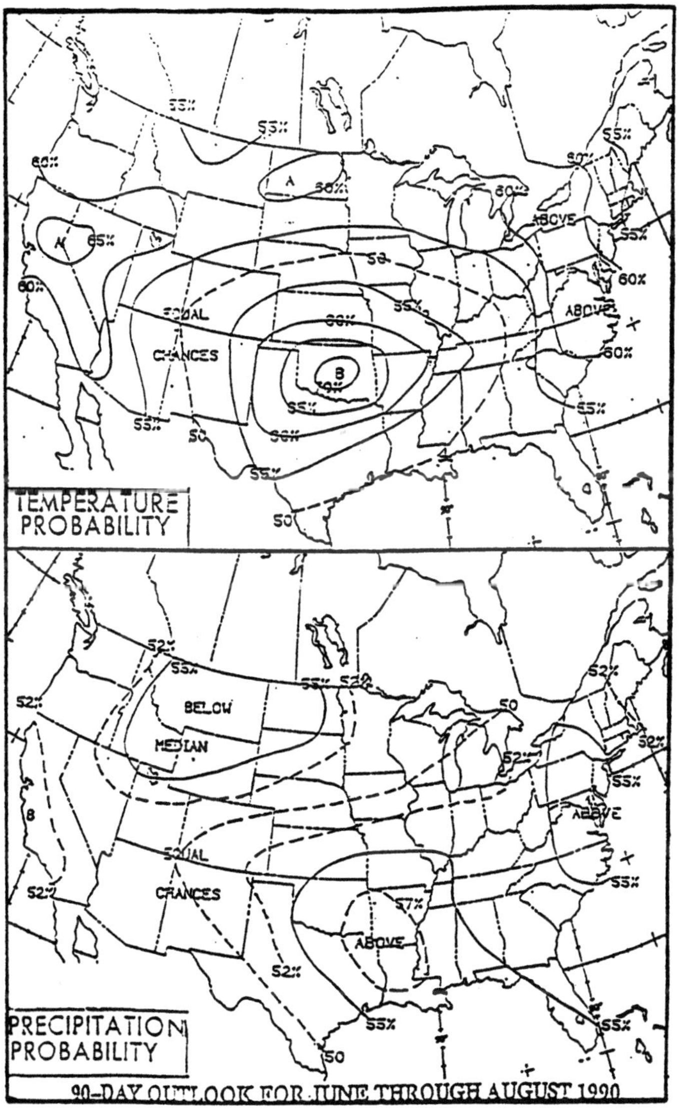

FIGURE 1: Probabilistic forecasts for JJA 1990.

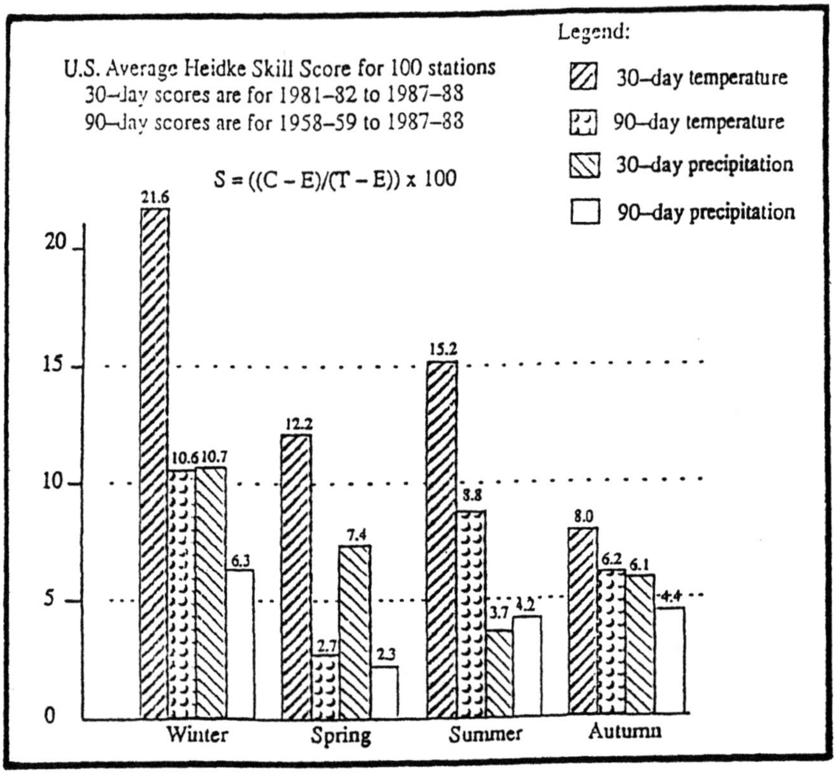

FIGURE 2: Relative accuracy of the operational monthly and seasonal forecasts stratified by season.

that P forecasts are less skillful than T, that skill is generally highest in winter and lowest in the transition seasons, and that monthly forecast skill exceeds seasonal skill. While the significance of the differences in skill among the various seasons is questionable, it is interesting to note that seasonal forecasts of T and P appear considerably less skillful in spring than in any other season. It has been observed that the predictability of ENSO is minimal through the Northern Hemisphere spring season (see, for example, Latif and Graham, 1991), and one might speculate on the possible connection between these two facts.

4. CONCLUDING REMARKS

This author (who is not involved in the seasonal prediction process) was struck by the similarity between the operational seasonal prediction scheme and the process used to produce short-range weather forecasts prior to the development of skillful numerical weather prediction forecast models. The obvious reason for this is that the current scheme strongly resembles the original one developed in the 1940's for long-range prediction, which was based on the short-range forecast process of that time. A more subtle reason, however, is that the process works - it results in forecasts which are, on average, more skillful than those produced by other methods that have been tried.

While improvements to the forecast process have been made - the mechanics of generating the prediction maps and issuing the forecasts have been automated to a considerable degree in the last 5 years - fundamental changes will probably have to await a development analogous to the arrival of useful numerical weather prediction models in the short-range prediction effort. For example, the demonstration of a coupled ocean-atmosphere model capable of predicting ENSO (something that may be near at hand - see article by Cane in this volume) would certainly result in changes to the existing operational scheme. Were the model to be a complex coupled general circulation model, the changes might prove as revolutionary as those that occured in short-range weather prediction during the 1960's and 1970's.

5. ACKNOWLEDGEMENTS

This paper could not have been written without the help of Drs. Huug van den Dool and Robert Livezey, both of whom provided advice and material before the oral presentation as well as comments on the manuscript. I should also thank Drs. Jerome Namias and Donald Gilman, who steered the operational process until recent years, and all the current practioners in the Prediction Branch of the Climate Analysis Center.

6. REFERENCES

Barnston AG, Livezey RE (1989) An operational multifield analog/anti-analog prediction system for United States seasonal temperatures. Part II: Spring, Summer, Fall and intermediate three-month period experiments. J Climate 2:513-541

Latif M, Grahm NE (1991) How much predictive skill is contained in the thermal structure of an OGCM. TOGA Notes 2:6-8

Livezey RE, Barnston AG (1989) An operational multifield analog/anti-analog prediction system for United States seasonal temperatures. Part I: System design and winter experiments. J Geophys Res 93:10 953-10 974

Livezey RE (1990) Variability of skill of long-range forecasts and implications for their use and value. Bull Am Meteor Soc 71:300-309

Namias J (1964) A 5-year experiment in the prepartion of seasonal outlooks. Mon Wea Rev 92:449-464

Ropelewski CF, Halpert MS (1986) North American precipitation and temperature patterns associated with the El Nino/Southern Oscillation (ENSO). Mon Wea Rev 14:2352-2362

Ropelewski CF, Halpert MS (1989) Precipitation patterns associated with the high index phase of the Southern Oscillation. J Climate 2:268-284

van den Dool HM (1988) A utilitarian atlas of monthly and seasonal precipitation-temperature relations, the United States, 1931-1987. University of Maryland, Department of Meteorology, SR-88-17, 36 pp

van den Dool HM, Nap JL (1985) Short and long range air temperature forecasts near an ocean. Mon Wea Rev 113:878-886

van den Dool HM, Klein WH, Walsh JE (1986) The geographical distribution and seasonality of persistence in monthly mean air temperature over the United States. Mon Wea Rev 114:546-560

Wagner AJ (1989) Medium- and long-range forecasting. Weather and Forecasting 4:413-426

EXPERIMENTAL SEASONAL FORECASTING OF TROPICAL RAINFALL AT THE UK METEOROLOGICAL OFFICE

M. N. Ward, C. K. Folland, K. Maskell, A. W. Colman, D. P. Rowell and K. B. Lane[1]
Hadley Centre for Climate Prediction and Research,
Meteorological Office, Bracknell, Berks,
RG12 2SY, England.

1. Introduction

Research in several centres suggests that sea-surface temperature anomalies (SSTAs) have a strong influence on seasonal circulation in the tropics (e.g. Nicholls 1987; Shukla and Fennessy 1988; Folland et al. 1991; Ward and Folland 1991). Motivated by this suggestion, real-time experimental forecasts of seasonal rainfall in the Sahel region of North Africa have been made by the UK Met. Office since 1986, and for the north Nordeste region of Brazil since 1987, a total of eleven such forecasts. The forecasts use the SSTAs 1-3 months before the main rainfall season to predict the rainfall total for that season. This paper first discusses the nature of the links between SSTAs and rainfall in the two forecast regions (section 2). Next we discuss the methods used to make the rainfall forecasts, and the amount of skill that the methods show on historical data (section 3). Forecasts that are made on historical data for the purpose of evaluating the forecast methods are referred to as *hindcasts*; the term *forecast* is reserved for the issued real-time forecasts. Section 4 discusses issues regarding the production of forecasts in real-time, and verifies eleven real-time seasonal rainfall forecasts. A forecast may fail if the large-scale SSTA patterns change significantly from the time the forecast is made to the time of the peak rainfall season. Section 5 therefore discusses preliminary work aimed at statistically predicting the evolution of SSTAs over a few months.

2. Relationships between sea-surface temperatures and seasonal rainfall in the Sahel and the North Nordeste

The Sahel (Fig. 1a) and the north Nordeste (Fig. 2a) both have a distinct rainfall season (mainly July to September in the Sahel and March to May in the north Nordeste). There is relatively little rainfall during the rest of the year in the Sahel, though rather more in the north Nordeste. This seasonality is largely governed by the north/south migration of the intertropical convergence zone.

Seasonal rainfall totals in the Sahel vary greatly on all timescales, including the interdecadal (Fig. 1b), while variations in the north Nordeste have been mostly confined to shorter period fluctuations through this century (Fig. 2b). The rainfall variability in both regions has been related to variations in SSTAs (e.g.

---

[1]Centre for Research in Environmental Systems, University of Lancaster, England, LA1 4YQ

NATO ASI Series, Vol. I 6
Prediction of Interannual Climate Variations
Edited by J. Shukla

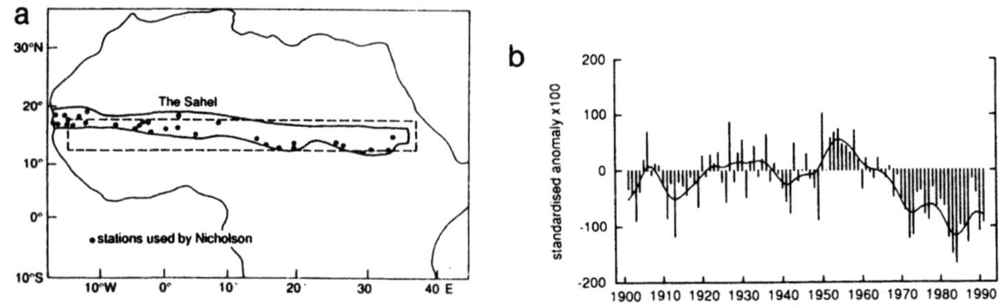

FIG. 1. (a) The boundary of the Sahel using Nicholson's (1985) definition (solid line, mean annual rainfall approximately 100 mm - 400 mm) and that appropriate to the GCM (dashed lines). (b) Standardised annual rainfall anomalies for the Sahel, 1901-91. Values to 1984 are from Nicholson (1985); 1985-91 values are based on CLIMAT or synoptic reports. Low-frequency component (solid line) fitted using Integrated Random Walk smoothing algorithm described in Ng and Young (1990) and Young et al. (1991) with 50% cut-off frequency set to about 10 years.

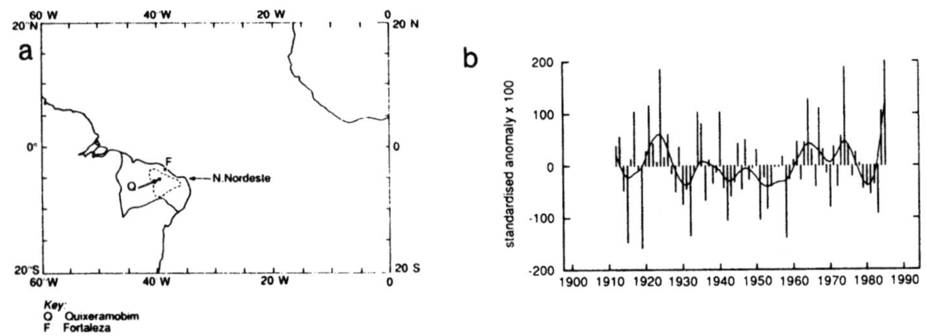

FIG. 2. (a) Location map of north Nordeste. The rainfall series of Nobre is based on stations throughout north Nordeste. That of Hastenrath et al. (1984) used stations from within the area contained by the dashed line. (b) Standardised seasonal rainfall anomalies for the north Nordeste, 1912-85. Values to 1981 for February to May from Nobre (INPE, pers. comm.); 1982-85 values are for March to April from Hastenrath (Univ. of Wisconsin, pers. comm.). Trend component as in Fig. 1b.

Markham and McLain 1977; Lamb 1978; Moura and Shukla 1981; Hastenrath et al. 1984; Folland et al. 1986; Palmer 1986; Semazzi et al. 1988; Druyan and Hastenrath 1991). A simplified summary of the relationships is provided by maps of the correlation between local SSTAs and the rainfall series (Fig. 3a-b). Unrotated and rotated covariance eigenvectors (EOFs) of SSTAs have been calculated to (a) identify some of the principal modes of variation of SSTAs, and (b) use the time coefficients of a few of the EOFs to succinctly express many of the relationships apparent in Fig. 3. In section 3 the time coefficients of these EOFs, measured prior to the rainfall season, are used to form predictors. In this section, simultaneous relationships are discussed. Some of the EOF patterns calculated to study north Nordeste rainfall are derived using just data for Dec-Feb. Note that once the EOF patterns have been derived, time coefficients (i.e. the time series) of the patterns can be calculated using any given set of SSTA fields (e.g. March-May).

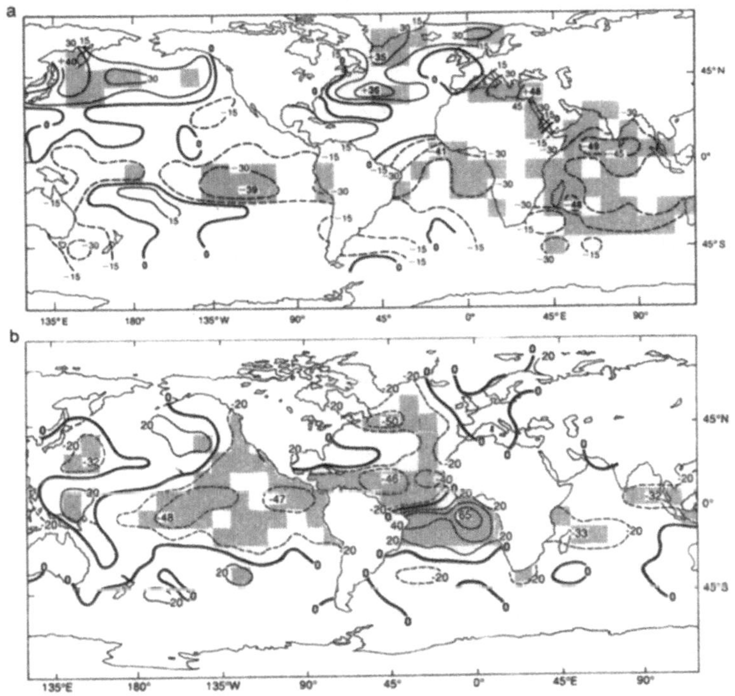

FIG 3. (a) Correlations (in hundredths) between 10° x 10° sea-surface temperature anomalies for June-August and standardised Sahel rainfall index for 1901-87. Correlations significant at the 5% level are shaded. (b) Correlations (in hundredths) between 10° x 10° sea-surface temperature anomalies for March-May and standardised north Nordeste rainfall index for 1912-85. Correlations significant at the 5% level are shaded.

2.1 Statistical relationships with EOFs

a. The North Nordeste

The first Pacific Dec-Feb EOF for 1901-80 (P1) is shown in Fig. 4 and the first three Atlantic all-seasons EOFs for 1901-80 (A1, A2, A3) are shown in Fig. 5 (Ward and Folland (1991) explain the selection of Dec-Feb data for P1 and all-seasons data for A1, A2 and A3). It has been found best to use only the tropical parts of the Atlantic EOFs as predictor patterns for the north Nordeste (details in Ward and Folland 1991). Time coefficients based on the tropical parts only are used throughout this paper. The tropical parts of Atlantic EOF n are referenced as ATn. EOFs AT3 and P1 have the strongest simultaneous correlations with north Nordeste rainfall (Table 1). P1 essentially represents the pattern of SSTA variability associated with the El Nino/Southern Oscillation phenomenon (ENSO). It is less influential than the Atlantic EOF, but is still important. This significant but not dominant role of ENSO is consistent with Ropelewski and Halpert (1987).

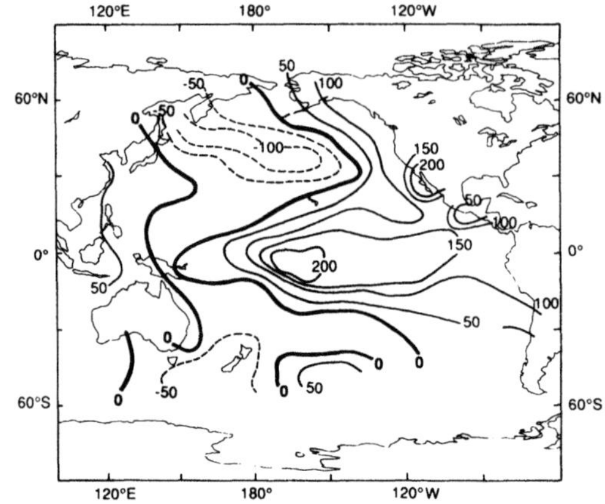

FIG. 4. First covariance eigenvector (EOF) of seasonal 10° x 10° Dec-Feb SST anomalies in the Pacific Ocean, 1901-80. Values are x1000. Anomalies are from the period 1951-80.

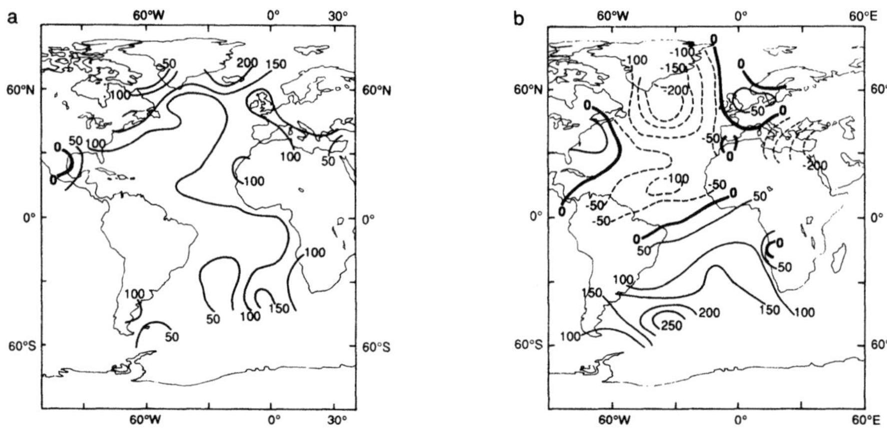

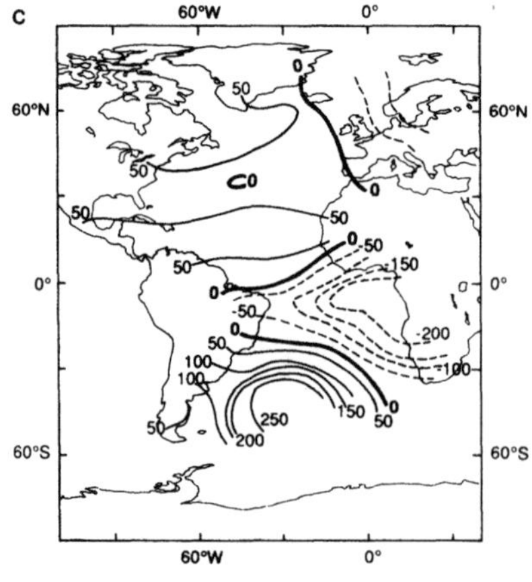

FIG. 5. Covariance eigenvectors (EOFs) of seasonal 10° x 10° SST anomalies in the Atlantic Ocean (all seasons), 1901-80. Values are x1000. Anomalies are from the period 1951-80. (a) EOF1, (b) EOF2, (c) EOF3.

When the EOF patterns are recalculated using data for a different period of years, different EOF patterns may be found, perhaps caused by:

(i) Uniqueness of SSTA patterns in one of the analysis periods.

(ii) Changes in the relative importance of SSTA patterns or changes in the overall noise level, leading to some SSTA variability being confounded among different EOFs due to statistical sampling.

Table 1. Correlation between north Nordeste rainfall index and SST eigenvector (EOF) time coefficients from 1912 to 1985.

	Data used for time coefficients	
	Mar to May	Nov to Jan
AT3	-0.62	-0.59
P1	-0.40	-0.29

Notes:
AT3 = third Atlantic EOF (tropical part)
P1 = first Pacific EOF

When the Atlantic all-seasons EOFs are recalculated for 1949-88, the latter process seems to have operated, since aspects of 1901-80 A2 and A3 can be seen in 1949-88 A1 and A4 (Fig. 6). Most importantly, the tropical part of 1901-80 A3 has been divided between 1949-88 A1 and A4 (details in Ward and Folland 1991). The consequences of this for our forecasting methods are discussed in section 3.1a.

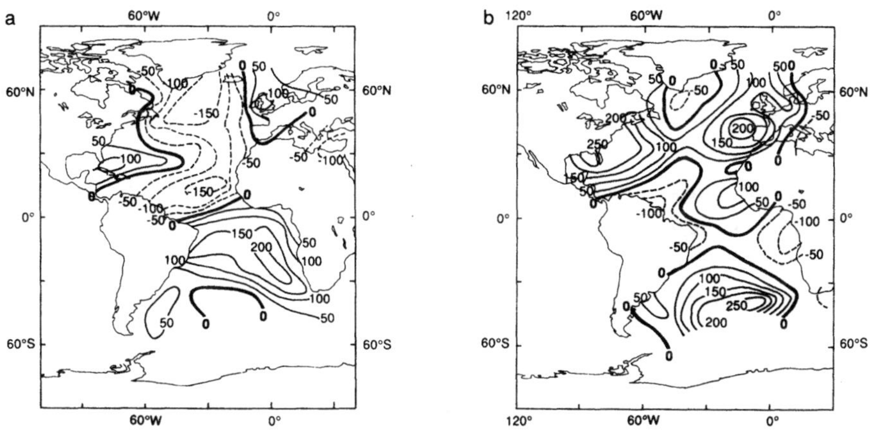

FIG. 6. Covariance eigenvectors (EOFs) of seasonal 10° x 10° Dec-Feb SST anomalies in the Atlantic Ocean, 1949-88. Values are x1000. Anomalies are from the period 1951-80. (a) EOF1, (b) EOF4.

b. The Sahel

Three 1901-80 global, all-seasons SSTA EOFs (Folland et al. 1991) are related to Sahel rainfall (see Fig. 7 and Table 2). The first EOF (EOF1, not shown) has weights of the same sign over most of the globe and represents global mean SSTA. It is not related to Sahel rainfall variability in this century. It has been found necessary to calculate EOFs using data for much of this century to ensure that the effects of global mean SSTA change are isolated in just one EOF, and not confounded with the other EOFs that represent patterns of SSTA variability that are important for Sahel rainfall.

The third global unrotated EOF (EOF3) (Fig. 7b) is a large-scale pattern with generally opposite weights in the northern and southern hemispheres (the northern Indian Ocean behaves like the southern hemisphere). The EOF time coefficients in July to September (JAS) are strongly correlated with Sahel rainfall. The Sahel tends to be wet when the southern hemisphere is cold and the northern hemisphere warm. Cross spectral analysis indicates that the association is strongest on decadal timescales (Folland et al. 1991) but also significant on some shorter timescales.

There are no convincing global-scale EOF patterns beyond EOF3 using this form of analysis so EOFs4-13 have been rotated using a VARIMAX rotation to bring out regional patterns. A scree test indicates that unrotated global EOFs below EOF13 represent noise. The second global rotated EOF (REOF2) is the only rotated global EOF to have a strong relationship with Sahel rainfall (Table 2). It consists mainly of a cross equator dipole in the tropical Atlantic with largest weights in the South Atlantic (Fig. 7c). Lamb (1978) and Lough (1986) found a similar pattern to be related to rainfall in the Sahel. Cross spectral analysis shows that the association with Sahel rainfall is strongest on shorter than decadal timescales.

There is a very weak correlation between drought in the Sahel and ENSO, represented by the second global unrotated EOF (EOF2) (Fig. 7a, Table 2). The association is strongest on timescales of 5 to 6 years (consistent with the period of the ENSO cycle) and around 2 years (Folland et al. 1991 give details). Using this simple measure of ENSO probably leads to an underestimate of the importance of the tropical Pacific for rainfall variations in the Sahel (Ward 1992), and work is underway to find a better predictor of Sahel rainfall using boreal spring ocean-atmosphere patterns in the western tropical Pacific.

Table 2. Correlation between Sahel rainfall index and global SST eigenvector time coefficients from 1901 to 1990.

	Data used for coefficients	
	Jul to Sep	Mar to May
EOF2	-0.25	-0.14
EOF3	-0.63	-0.56
REOF2	-0.57	-0.39

Notes:
EOF2 = second unrotated EOF
EOF3 = third unrotated EOF
REOF2 = second rotated EOF (see Fig. 7)

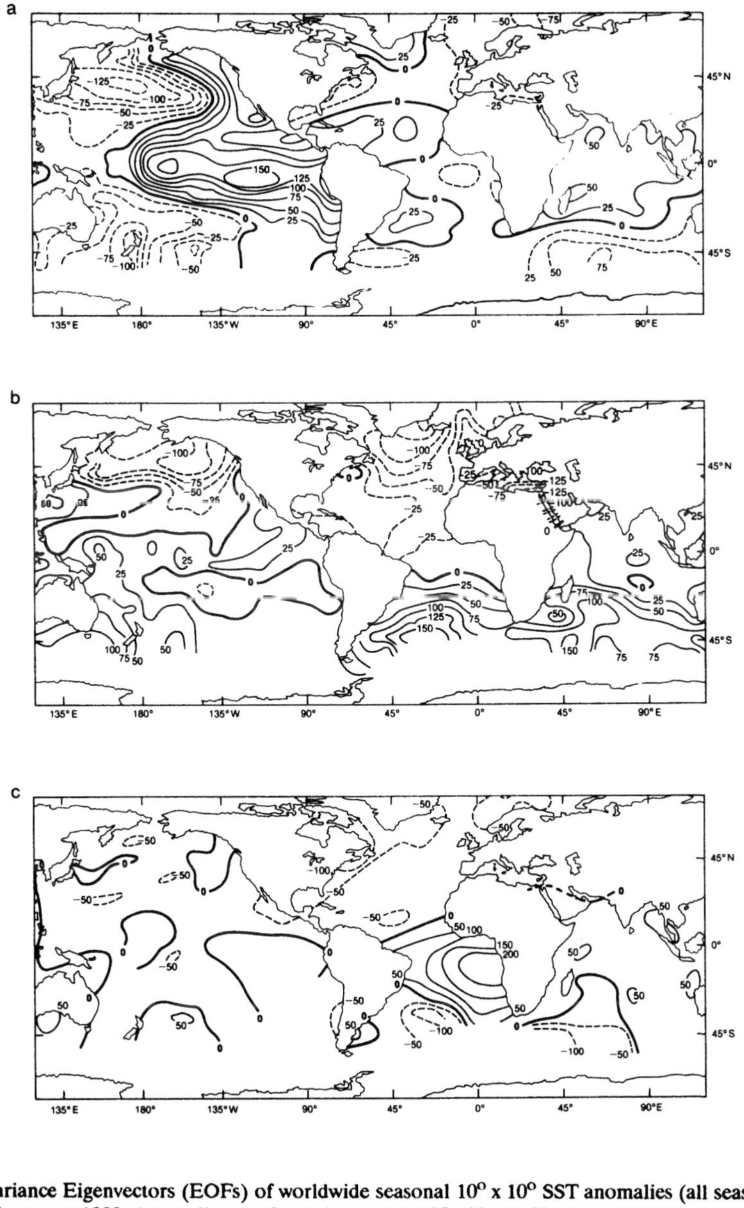

FIG. 7. Covariance Eigenvectors (EOFs) of worldwide seasonal $10^o \times 10^o$ SST anomalies (all seasons) for 1901-80. Values are x1000. Anomalies are from the period 1951-80. (a) Unrotated EOF2, (b) Unrotated EOF3, (c) Rotated EOF2 (derived by VARIMAX rotation of EOFs 4-13).

2.2 General circulation modelling results

Evidence for the influence of SSTAs on Sahel rainfall has been greatly strengthened by a series of experiments with a version of the UK Met. Office general circulation model (GCM) forced with observed SSTs from 10 past years following preliminary studies by Folland et al. (1986) and Palmer (1986). The years selected represent a broad cross-section of variability in the Sahel: 1949, 1950, 1958, 1976, 1983, 1984, 1987, 1988, 1989 and 1990. The GCM has 11 vertical sigma levels and a horizontal resolution of 2.5°latitude by 3.75°longitude which gives 28 model grid points in the Sahel (14 in longitude, 2 in latitude). These points cannot be made to coincide with Nicholson's (1985) area exactly; the effective area covered by the GCM is shown in Fig. 1a. The GCM area does not include the extreme western Sahel (this point is 'sea' at the model resolution) and on average is further south than Nicholson's Sahel and therefore covers part of Nicholson's 'Soudan' region. The GCM has a reasonable rainfall climatology in the Sahel, both temporally and spatially, during the main part of the wet season (JAS) (Rowell et al. 1991). Each experiment started from initial atmospheric data for the end of March and ran until the end of October (two sets of initial atmospheric data were used as a test of the model's sensitivity to initial conditions; 26-3-84 and 31-3-87). The SST data were taken from the Met. Office historical SST (MOHSST) dataset (Bottomley et al. 1990). The SST values were updated every 5 days throughout the integration. In regions of no observations the SST data were filled in using an adjustment to a background 1951-80 climatological data set using Poisson's equation (Reynolds 1988). For a more detailed description of experimental design and results see Folland et al. (1991) and Rowell et al. (1991). Early results are discussed in Owen and Folland (1988) and Folland et al. (1989). Such simulations of Sahel rainfall can be thought of as a test of the GCM's predictive ability given perfect knowledge of the SST evolution from April to October.

Figure 8 illustrates the capability of the model to simulate with reasonable accuracy the seasonally averaged rainfall anomaly for most of the 10 selected years (the main exception being 1990). The high correlation of observed and simulated JAS rainfall (Table 3) suggests that SSTAs are responsible for a considerable proportion of interannual seasonal rainfall variability in the Sahel. Perhaps a better estimate of this proportion is given by the absolute error expressed as a fraction of the chance value (Table 3). Skill is only modestly reduced when the Sahel is divided equally into West and East (Rowell et al. 1991).

3. Prediction techniques

Large-scale SSTA patterns often change fairly slowly on the timescale of a season and can even persist for a few seasons or longer. Time coefficients of the EOFs discussed in section 2 have been created using SSTAs observed prior to the rainfall season. The correlation of these coefficients with the seasonal rainfall indices is lower but still statistically significant (except for EOF2 v Sahel rainfall) (see Tables 1 and 2). This suggests that SSTA patterns measured prior to the main rainfall season have some potential for skilful prediction of Sahel and north Nordeste rainfall. Both statistical and GCM techniques are discussed below.

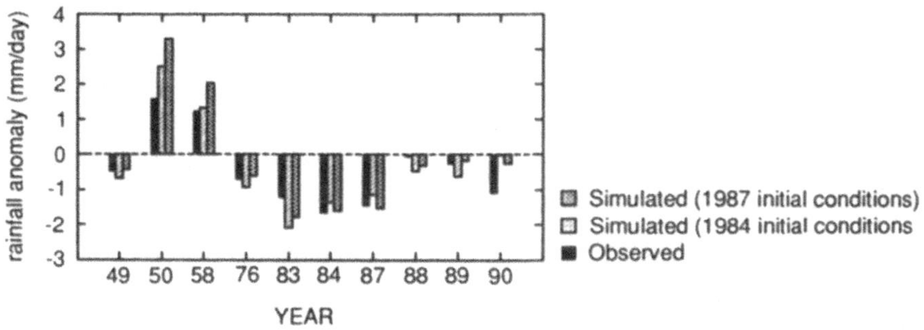

FIG. 8. Histogram comparing the observed and the GCM simulated July to September rainfall anomalies for 10 past years. Each integration of the GCM was forced with the SST observed in the given year. The integrations were run starting from two different initial atmospheric conditions. The zero line is the observed 1951-80 climatology (3.36 mm day^{-1}). Observed rainfall values interpolated to the GCM grid were kindly provided by Dr M Hulme, Climatic Research Unit, Norwich, UK under DoE PECD 7/12/78.

Table 3. GCM simulations of July-September Sahel rainfall in 10 past years (see Fig. 8)

Initial atmospheric data	r	bias (mm/day)	ABSE (mm/day)	% of chance ABSE
26-3-84	0.90	0.07	0.47	37
31-3-87	0.94	0.29	0.45	32

r = Correlation (standard type) between the simulated and observed seasonal rainfall anomaly.
ABSE = Mean absolute error of the simulated seasonal rainfall anomaly.
chance ABSE = ABSE when simulated values are randomly re-arranged.

3.1 Statistical prediction techniques

Two statistical techniques have been used; multiple linear regression (MLR) and linear discriminant analysis (LDA) (see Ward and Folland 1991 for a full discussion of their application). The predictors for both techniques are time coefficients of the SSTA EOFs discussed in section 2.1. MLR predicts a single best estimate value of the standardized rainfall anomaly while LDA predicts the probabilities that the rainfall will be in each of 5 categories or quints. The quints are calculated to have an equal chance of occurrence over the total training period of the model. Explicit forecasts of the posterior probability of the rainfall categories are a very useful feature of LDA. It can be argued that the technique is in better accord with the probabilistic nature of long-range forecasting than is multiple regression. Ward and Folland (1991) illustrate that for the north Nordeste hindcasts 1912-85, the probabilities estimated by LDA for each category are giving fairly accurate estimates of the likely occurrence of the categories i.e. the probability forecasts are well calibrated. The techniques have been tested extensively for both regions using independent training periods (over which the models are developed) and testing (or hindcasting) periods.

a. The North Nordeste

Table 4 assesses MLR and LDA hindcasts for 1912-48 and 1949-85 that have been made using models with training periods 1949-85 and 1912-48 respectively. Seasonally (three month) averaged values of the predictor EOF time series are used to minimise the noise in the SSTA data that is likely to be a problem in earlier decades. In Table 4, only two predictors are used, time coefficients of 1901-80 AT3 and P1. These models correspond to those selected by a stepwise process with a significance level of 5% in 1912-48 and 1% in 1949-85. Results from models using a 10% significance threshold were generally inferior.

Table 4. Assessment of multiple linear regression (MLR) and linear discriminant analysis (LDA) hindcasts of north Nordeste rainfall based on 1901-80 EOFs.

			MLR hindcasts				LDA hindcasts	
hindcast period	training period	hindcast from	r_f	r_p	hits skill	LEPS CAT	hits skill	LEPS CAT
1949-85	1912-48	Nov-Jan	.74	.24	22%	53%	25%	54%
1949-85	1912-48	Dec-Feb	.76	.24	38%	61%		
1912-48	1949-85	Nov-Jan	.63	.11	24%	38%	25%	31%
1912-48	1949-86	Dec-Feb	.76	.11	37%	51%	46%	54%

Notes:
r_f = linear correlation between forecast and observed standardised seasonal rainfall anomaly
r_p = linear correlation achieved by persistence forecast strategy
hits skill = assessment of number of forecasts that reside in the correct quint
LEPS CAT = Linear Error in Probability Space, assessment of quint forecasts
For hits skill and LEPS CAT: the score achieved on average by a set of random forecasts is 0% and the score achieved by a set of perfect forecasts is 100% (details in Ward and Folland 1991).

All models perform much better than persistence. Bias values (not shown) are generally about one standard error from a zero value, so are not statistically significant. The correlations between hindcast and observed rainfall suggest about 55% of north Nordeste seasonal rainfall variance may be predictable from SSTAs measured over the period December - February. Using SSTA data to the end of January, the hindcast skill is a little less in the period 1912-48, but again close to 55% for the more recent period 1949-85.

It can be argued that the above hindcast periods were not completely independent of model construction periods, because the EOF patterns were based on 1901-80 data (even though the time coefficients of the EOFs were measured independently in the 1912-48 and 1949-85 training and testing periods) (e.g. Hastenrath, 1990 p903). To assess the impact of this on hindcast skill, the 1949-88 EOFs are used as candidate predictors in a 1949-85 training period, and hindcasts are made for the fully independent period 1912-48.

The three variables AT1, AT4 (Fig. 6) and P1 (not shown; the 1949-88 version of P1 is very similar to the 1901-80 version in Fig. 4) combine well as candidate predictors, all entering the regression model at a significance <1%. The skill of the 1912-48 hindcasts using these three EOFs (Table 5) is remarkably similar to that achieved using 1901-80 EOFs. For more details see Ward and Folland (1991).

Table 5. Assessment of MLR hindcasts of north Nordeste rainfall based on 1949-88 EOFs.

hindcast period	training period	hindcast from	r_f	r_p	hits skill	LEPS CAT
1912-48	1949-85	Nov-Jan	.60	.11	18%	37%
1912-48	1949-85	Dec-Feb	.75	.11	27%	45%

For skill score definitions see Table 4.

b. The Sahel

Table 6 assesses MLR hindcasts for 1901-45 and 1946-90 that have been made using models created for the training periods 1946-90 and 1901-45 respectively (LDA hindcasts are in Folland et al. 1991). The prediction techniques have skill well above the chance level. The predictors used for the results in Table 6 are EOF2, EOF3 and REOF2. The 1901-45 hindcasts score significantly better than persistence. In 1946-90, persistence is much higher and is difficult to beat. However, note that the MLR model that makes the skilful hindcasts for 1946-90 was trained on a highly non-persistent period (1901-45).

Table 6. Assessment of MLR hindcasts of Sahel rainfall based on 1901-80 EOFs.

hindcast period	training period	hindcast from	r_f	r_p	hits skill	LEPS CAT
1901-45	1946-90	Apr-May	.51	.15	23%	39%
1946-90	1901-45	Apr-May	.73	.67	18%	48%
1946-90*	1901-45	Apr-May	.73	.67	30%	70%

* These 1946-90 hindcasts produced using inflated regression.

For skill score definitions see Table 4.

3.2 Prediction using a GCM

The potential of the GCM described in section 2.2 for Sahel rainfall prediction has also been investigated. The model was forced through the Sahel rainfall season with SSTs calculated by persisting the anomalies from each of the months April, May and June. The persisted anomalies were added to an evolving SST 1951-80 climatology. This is one simple way to 'predict' the summer SSTs from spring SSTs. The method has been tested on 7 years (Fig. 9 and Table 7). A high level of skill is achieved when SSTAs are fixed from June, the only failure being in 1990. In real-time, a forecast using June SSTAs could not be issued until early July (just after the start of the main rainfall season); nevertheless such a forecast can still be of practical use. Skill drops substantially when April or May SSTAs are used. Based on this limited sample of years, the skill of the GCM technique seems to be less than that of the statistical techniques when the forecast lead-time is more than a month. The results suggest that simple persistence of SSTAs may be inadequate for use with the GCM except in the "nowcasting" situation of an updated forecast in early July. If accurate predictions of boreal summer SSTAs could be produced in boreal spring, the GCM would become a more powerful forecasting tool. SSTA prediction is discussed in section 5.

Table 7. GCM hindcasts of July-September Sahel rainfall in seven past years (see Fig. 9)

hindcast from	r	bias (mm/day)	ABSE (mm/day)	% of chance ABSE
June	0.88	0.23	0.64	37
May	0.37	0.32	1.16	69
April	0.30	-0.32	1.10	86

For skill score definitions see Table 3.

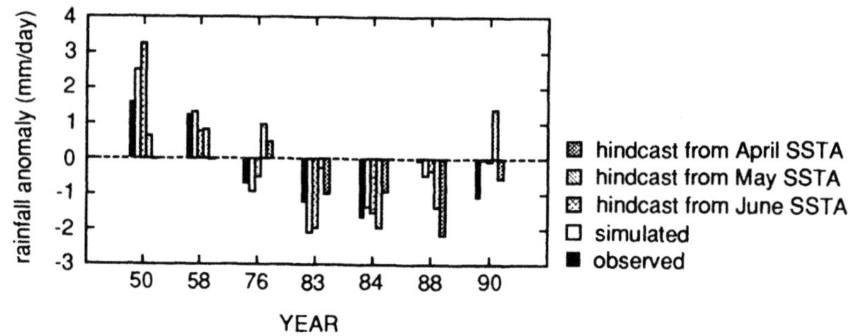

FIG. 9. Histogram of the observed, GCM simulated (1984 initial data) and GCM hindcast July to September Sahel rainfall anomalies for 7 past years, including real-time experimental forecasts for 1990. The zero line is the observed 1951-80 climatology (3.36 mm day^{-1}). From left to right, the bars are for the observed, simulated, hindcast from June SSTA, hindcast from May SSTA and hindcast from April SSTA.

4. Experimental real-time forecasts

a. North Nordeste forecasts

These forecasts have been made by up-dating the 1901-80 SSTA EOF time coefficients in real-time, and applying them to statistical prediction models. Forecasts using models from a number of training periods are calculated, and the forecasts averaged for the final product. Preliminary forecasts are made from SSTA data up to the end of January, followed by a final forecast including SSTA data for February. Experimental forecasts have been made for the average of the standardised March-May rainfall totals at Quixeramobim and Fortaleza (QF index) (see Ward et al. 1988). Forecasts have also been made for Hastenrath's March-April index and, since 1988, for Nobre's February-May index. Only the QF index has been updated, so we use this index to verify the forecasts.

The forecasts are distributed in early March each year to interested scientists and scientific institutions. Since 1990, the preliminary forecasts have been made available to FUNCEME (Meteorological Centre of Ceara State in the Nordeste) in early February, so that they can contribute to the monitoring and forecast service that FUNCEME now provides for the region.

Forecasts are made using both of the statistical techniques (MLR and LDA) that were discussed in section 3.1. Fig. 10a verifies the preliminary and updated forecasts that were issued prior to the five wet seasons 1987-91. In Fig. 10a the forecasts are expressed in terms of the five equiprobable rainfall categories that we have used. The forecast category shown is based on the average of all the forecasts that were made. The skill is generally very good. An important issue concerns the averaging period for the SSTA data that is used to make the forecasts. Generally, hindcast skill is best when using a three month averaging period. However, this may not be true for all years, especially when the SSTA field is changing rapidly. In such instances, the average of the latest two or even one month's SSTAs may provide the optimum forecast. This strategy has been subjectively applied when producing the forecasts.

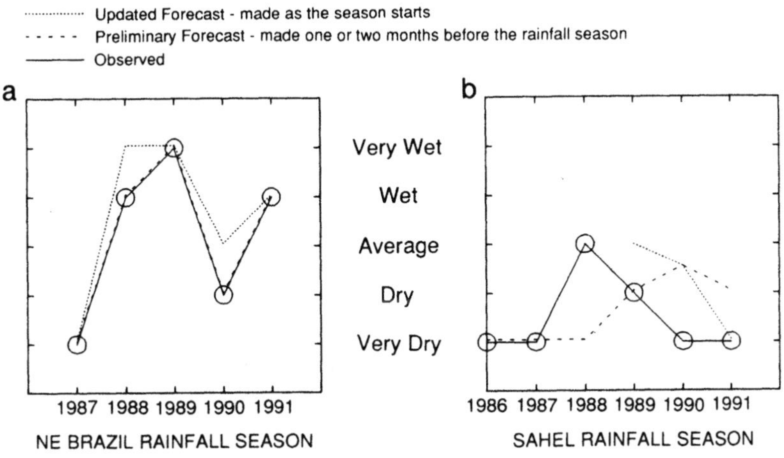

FIG. 10. Experimental real-time forecasts of (a) north Nordeste and (b) Sahel seasonal rainfall, in terms of rainfall quints that are equiprobable in the period 1951-80 for north Nordeste and 1941-85 for the Sahel. North Nordeste observed is provisional based on March-May series for Quixeramobim/Fortaleza (Ward et al. 1988). The Sahel 1990 forecasts made quints 2 and 3 equally most likely.

b. Sahel forecasts

Statistical forecasts using EOFs are made for the Sahel in a similar way to those for the north Nordeste. Since 1988, the GCM forecast method has also been used. Prior to 1989, the last released forecast used SSTA data up to the end of May. However, an investigation of the forecast failure in 1988 highlighted the large improvement possible by using June SSTA data (Fig. 9). So since 1989 a pattern has been established of a preliminary forecast based on data up to the end of April, followed by an updated forecast based on data to the end of June. The forecasts are circulated to interested scientists, Heads of Sahelian National Meteorological Agencies and a few International Aid Agencies who have specifically requested the information.

Compared to the north Nordeste, the skill of the Sahel forecasts (Fig. 10b) is more variable, being good in 1986, 1987, 1989 and 1991 but poor in 1990 and especially 1988 when the forecast was issued for exceptionally dry conditions. However, this is only a small sample of years, and the hindcast results of section 3 suggest that over a longer period the level of skill in the two regions is likely to be similar.

5. Prediction of July-September sea-surface temperature anomalies

5.1 Introduction

The above forecast methods may produce a poor forecast when SSTAs change greatly from the time the forecast is made to the time of the peak rainfall season (as occurred from boreal spring to summer in 1988). The GCM is especially sensitive to such changes because it relies on simple persistence of SSTA. (The statistical models implicitly include the effect of typical SSTA changes but will fail when SSTA changes are markedly atypical). Figure 11 shows the persistence correlation of SSTA between the seasons March to May (MAM) and JAS. There are many regions where the correlation falls below 0.5 and only small regions where it is greater than 0.7.

Previous research into forecasting SSTAs has concentrated on predicting ENSO (Graham et al. 1987; Cane et al. 1986; also reviewed in this volume). The need here is to develop improved ways of predicting SSTAs in as many regions as possible. Three ways to improve upon simple SSTA persistence are introduced in the next section.

5.2 Improvements on persistence

a. Re-scaled persistence

Even in regions where the persistence correlation is high, persistence of SSTA from boreal spring to summer may not be appropriate due to large differences in the variance of SSTAs in MAM and JAS (Fig. 12). This is particularly important in regions of local SST maxima. For example, in the tropical North

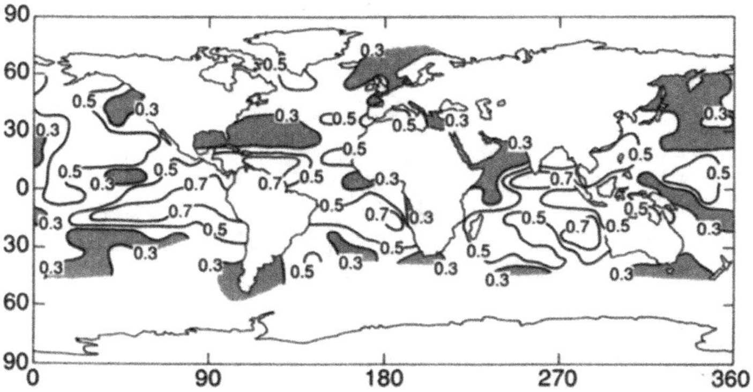

FIG. 11. Correlation (hundredths) between MAM SSTA and JAS SSTA for each 10° latitude x 10° longitude box over 1949-88. Contours at 0.3, 0.5, 0.7; values <0.3 are shaded.

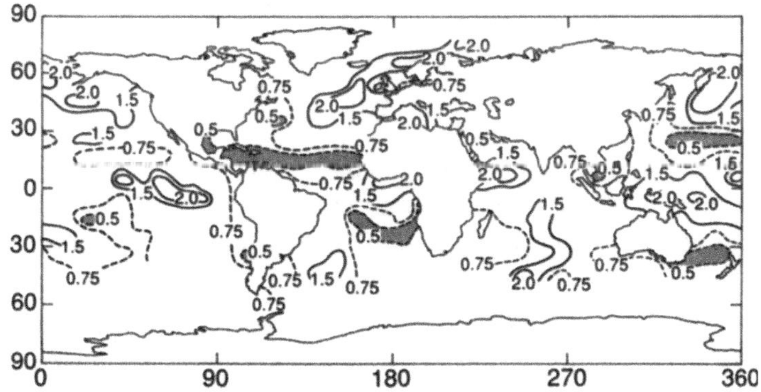

FIG. 12. Ratio of SSTA variance in JAS to SSTA variance in MAM. Using 1949-88 data. Dashed contours at 0.5, 0.75; solid contours at 1.5, 2.0; values <0.5 are shaded.

Atlantic at 10-20°N, the JAS SSTA variance is between 10-50% of the MAM variance. So persistence of MAM SSTA will in many years produce unreasonably large or small SST values, and most importantly, introduce unprecedented SST gradients. This may have severe consequences for the response of the GCM, since in regions of high SST in the tropics the surface wind convergence is often largely determined by the Laplacian of SST (Barnett et al. 1991). Another region where the variance ratio is likely to be important is the western tropical Pacific. A relatively low SST variance and the local SST maximum are found in MAM near to and just south of the equator, while in JAS, maximum SST and low SST variance are both found further north. Such problems may partially explain the poor performance of GCM hindcasts from April and May SSTAs (see section 3.2). Application of statistical regression models using the boreal spring SSTA to forecast the summer SSTA will ensure the forecast SSTAs are *re-scaled* to be of an appropriate magnitude. Ordinary regression models will produce forecasts with a standard deviation that is less than the observed by an amount that ensures the mean square error of the forecast is minimised. If the variance of the forecasts is required to be the same as the observed, then inflated regression models should be used (as discussed in the context of predicting north Nordeste rainfall by Ward and Folland (1991)).

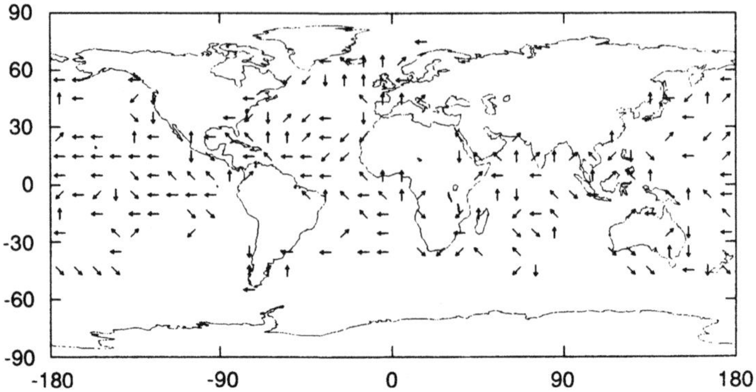

FIG. 13. Boxes (10° latitude x 10° longitude) for which JAS SSTA is better predicted by MAM SSTA in an adjacent box than by persistence. Arrows point away from the box with highest correlation. Correlations calculated for 1949-88.

b. Anomaly advection

Surface ocean currents can typically flow at around 10 cm s^{-1}, which implies advection of SSTAs of about 10° latitude in three months. So in some regions the JAS SSTA may be predicted more skilfully by the MAM SSTA one 10° x 10° box away, than by persistence. To investigate this the correlation between JAS SSTA in a given box, and MAM SSTA in surrounding boxes, has been calculated, and the values compared with the persistence correlation. Figure 13 indicates those boxes where the persistence correlation is improved upon by the correlation with a surrounding box, with an arrow plotted to indicate the location of the highest correlation. The persistence correlation is, of course, improved upon in some boxes merely due to sampling fluctuations. Confidence in an ability to beat persistence results therefore when the arrows in Fig. 13 show spatial coherence, especially when their direction is consistent with the climatological surface ocean current. Examples of this are found in a number of regions, such as the tropical North Atlantic.

c. Large-scale predictors of sea-surface temperature anomalies

The large-scale atmospheric and oceanic patterns in boreal spring contain the seeds of some of the SSTA evolution from spring to summer. EOFs of global SSTAs (discussed in previous sections) and VARIMAX rotated correlation EOFs of sea-level pressure (SLP) calculated for data north of 15°N are used to represent the large-scale climate state in boreal spring. (The SLP REOFs are calculated following the discussion by Barnston and Livezey (1987)). The analyses use only data sources that are available at the UK Met. Office in real-time, so that the results can be applied to produce SSTA forecasts in real-time. This precluded the use of promising analyses performed using COADS SLP and surface wind. Nonetheless, useful predictability has been discovered, especially in some regions of low SSTA persistence in the tropics.

a

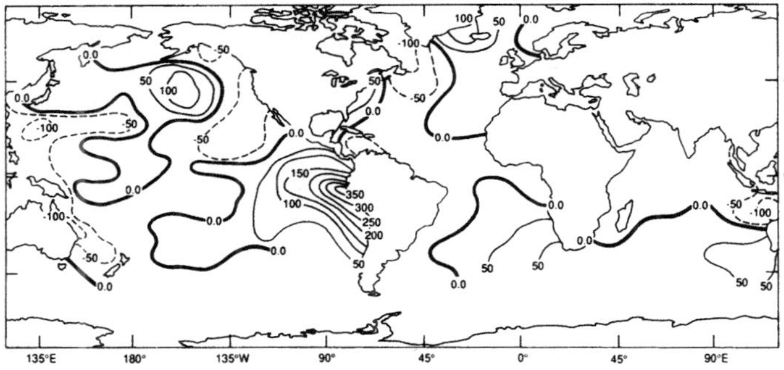

b

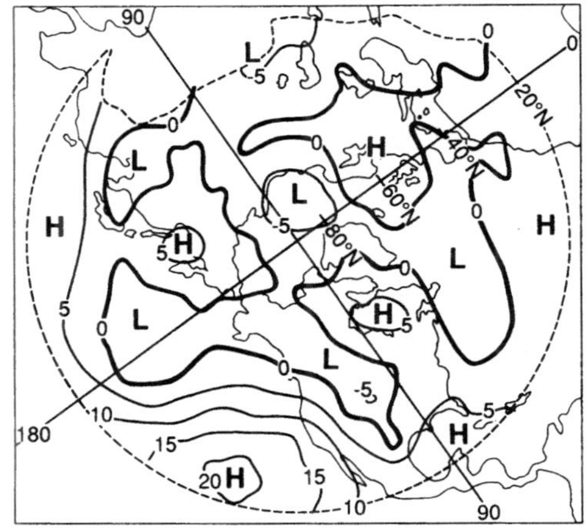

c

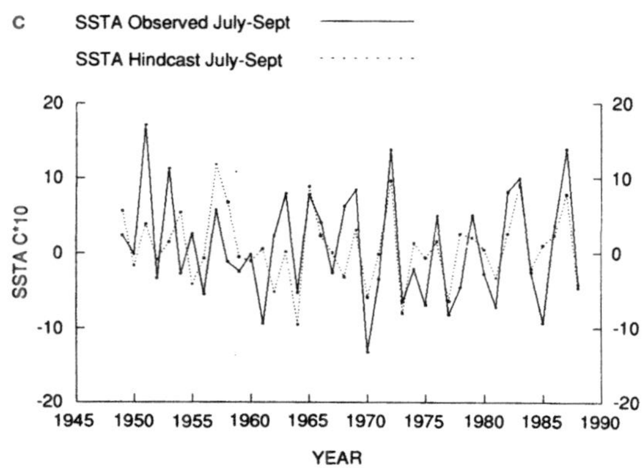

FIG. 14. (a) Fourth VARIMAX Rotated Covariance Eigenvector (SST REOF4) of seasonal 10° x 10° SSTAs (all seasons) for 1901-80. (b) Eighth VARIMAX Rotated Correlation Eigenvector (SLP REOF8) of half-monthly 5° latitude x 10° longitude SLP anomalies for May-June 1949-88 (using UK Met. Office data set). (c) Hindcast SSTA 1949-88 in 10° x 10° tropical Pacific box centred at 5°N 105°W. The predictors are the March-May time coefficients of the two EOFs in this Figure. The hindcasts are made from multiple regression models using the cross validation method to achieve a partly independent verification. Correlation = 0.61 (persistence correlation = 0.42).

To illustrate the types of relationships being investigated, Fig. 14 shows SLP REOF8 and SST REOF4 whose time series combine to produce a good regression forecast model for much of the equatorial central and eastern (ECE) Pacific Ocean. The MAM values of SLP REOF8 and SST REOF4 have near zero correlation (1949-88), so are explaining independent portions of SSTA variability in the ECE Pacific. SST REOF4 has large positive values close to Peru, similar to the pattern of warming in MAM which tends to develop in this region prior to a full-scale ENSO warm event (Halpert and Ropelewski 1989 Fig. 71b). The time coefficients of this pattern in MAM are positively correlated with ECE Pacific SSTAs in JAS. SLP REOF8 has positive weights in the subtropical northeastern Pacific and the time coefficients in MAM are negatively correlated with JAS SSTAs in the ECE Pacific. This implies an association between a stronger subtropical high and cooler ECE Pacific SSTAs, possibly as a result of stronger easterly trade winds leading to increased equatorial upwelling (Trenberth 1991). Using 1949-88 MAM values of these two EOFs, the multiple regression prediction model for the 10° latitude x 10° longitude JAS SSTA centred at 115°W, 5°N is (the EOF time series were standardised prior to fitting the model to have a standard deviation of 1 and a mean 0):

$$SSTA = 0.08 - 0.36 * SLP_{R8} + 0.30 * SST_{R4} \qquad (1)$$

The equation's coefficients for both EOF predictors are statistically significant at much better than 1%, and the multiple correlation coefficient for the model is 0.66, which compares with a persistence correlation of 0.42. Note that SST EOF2 (mature El Nino EOF, Fig. 7a) is not statistically significant if used in place of SST REOF4. Figure 14c shows the hindcast against the observed SSTA in this box using the cross validation method which provides a largely (not completely) independent test. The correlation between hindcast and observed (0.61) is only slightly less than the full dependent period correlation (0.66).

5.3 Forecasting worldwide sea-surface temperature anomaly patterns

No ideal statistical technique seems to exist to combine the variety of predictive information introduced above. The information is on a variety of spatial scales, and is contained in variables which sometimes are highly correlated. One approach is to calculate a regression forecast model for every 10° x 10° JAS SSTA box that contains data. For most boxes in the extratropics, the best predictor currently available is persistence, so the forecast is the boreal spring SSTA re-scaled by the regression coefficient (as discussed in section 5.2a). We call such a forecast *re-scaled persistence*. However, for most tropical boxes, models like the one shown in Eq. (1) perform better than re-scaled persistence. Usually, the best model includes one or two large-scale predictors (the EOFs) and a local predictor (either the boreal spring SSTA in the same box or an adjacent box). For example, in most central and eastern Pacific 10° x 10° boxes just south of the equator, the boreal spring SSTA to the east of the box being predicted (see Fig. 13) was added to the SLP REOF8 and SST REOF4 predictors to form the best model.

6. Conclusions

The seasonal rainfall forecasting systems for the tropics so far developed at the UK Meteorological Office are primarily based on SST information. Research elsewhere suggests that other sources of predictability exist. This is well illustrated by the recent work of Hastenrath (1990) on north Nordeste rainfall where pre-season rainfall is used successfully as a predictor. This is physically feasible since pre-season rainfall may modify the land surface characteristics such as soil moisture, which in turn could affect the evolution of the main rainfall season. However such additional predictors are likely to be correlated with SST predictors, so their inclusion into empirical forecast models will require care.

The current forecasts have a lead-time of at most two months on the start of the rainfall season. Even at this relatively short lead-time, substantial changes in large-scale SSTA patterns occasionally occur from the time the forecast is made to the time of the main rainfall season. This paper has reported some preliminary work on predicting SST anomalies a season in advance. Such work seems likely to improve forecast reliability at a lead-time of a few months.

REFERENCES

Barnett TP, Latif M, Kirk E, Roeckner E (1991) On ENSO Physics. J Climate 4:487-515.

Barnston AG, Livezey RE (1987) Classification, seasonality and persistence of low-frequency atmospheric circulation patterns. Mon Wea Rev 115:1083-1126.

Bottomley M, Folland CK, Hsiung J, Newell RE, Parker DE (1990) Global Ocean Surface Temperature Atlas (GOSTA). Joint Meteorological Office/Massachusetts Institute of Technology Project. Project supported by US Dept of Energy, US National Science Foundation and US Office of Naval Research. Publication funded by UK Depts of Energy and Environment. pp20+iv and 313 Plates. HMSO, London.

Cane MA, Zebiak SE, Dolan SC (1986) Experimental forecasts of El Nino. Nature 321:827-832.

Druyan LM, Hastenrath S (1991): Modelling the differential impact of 1984 and 1950 sea-surface temperatures on Sahel rainfall. Int J Climatol 11:367-380.

Folland CK, Palmer TN, Parker DE (1986) Sahel rainfall and worldwide sea temperatures, 1901-85, Nature 320:602-607.

Folland CK, Owen JA, Maskell K (1989) Physical causes and predictability of variations in seasonal rainfall over sub-Saharan Africa. In: Remote Sensing and Large-Scale Global Processes (Proceedings of the IAHS Third Int Assembly, Baltimore, MD, May 1989), IAHS Publ No 186, p87.

Folland CK, Owen JA, Ward MN, Colman AW (1991) Prediction of Seasonal Rainfall in the Sahel Region of Africa using empirical and dynamical methods. J Forecasting 10:21-56.

Graham NE, Michaelsen J, Barnett TP (1987) An Investigation of the El Nino-Southern Oscillation Cycle With Statistical Models. 2. Model Results. J Geophys Res, 92C13:14271-14289

Halpert MS, Ropelewski CF (1989) Atlas of tropical sea surface temperature and surface winds. NOAA Atlas No. 8, US Dept. Commerce, National Oceanic and Atmospheric Administration, National Weather Service, Silver Spring, MD.

Hastenrath S, Wu M-C, Chu P-S (1984) Towards the monitoring and prediction of north-east Brazil droughts. Q J R Met Soc 110:411-425.

Hastenrath S (1990) Prediction of Northeast Brazil rainfall anomalies. J Climate 3:893-904.

Lamb PJ (1978) Large-scale tropical Atlantic surface circulation patterns associated with Subsaharan weather anomalies. Tellus 30:240-251.

Lough JM (1986) Tropical Atlantic sea surface temperatures and rainfall variations in Subsaharan Africa. Mon Wea Rev 114:561-570.

Markham CG, McLain D (1977) Sea surface temperature related to rain in Ceara, north-eastern Brazil. Nature 265:320-323.

Moura AD, Shukla J (1981) On the dynamics of drought in Northeast Brazil: Observations, theory and numerical experiments with a general circulation model. J Atmos Sci 38:2653-2675.

Ng CN, Young PC (1990) Recursive Estimation and Forecasting of Non-stationary Time series. J Forecasting 9:173-204.

Nicholls N (1987) Prospects for drought prediction in Australia and Indonesia. In: Wilhite DA, Easterling WE, Wood DA (eds) Planning for Drought, Westview Press, Boulder and London, p61.

Nicholson SE (1985) Sub-Saharan rainfall 1981-84. J Clim Appl Met 24:1388-1391.

Owen JA, Folland CK (1988) Modelling the influence of sea-surface temperatures on tropical rainfall. In: Gregory S (ed) Recent Climate Change, Belhaven Press, London, p141.

Palmer TN (1986) Influence of the Atlantic, Pacific and Indian Oceans on Sahel rainfall. Nature 322:251-253.

Reynolds RW (1988) A real-time global sea surface temperature analysis. J Climate 1:75-86.

Ropelewski CF, Halpert MS (1987) Global and regional scale precipitation patterns associated with the El Nino/Southern Oscillation. Mon Wea Rev 115:1606-1626.

Rowell DP, Folland CK, Maskell K, Owen JA, Ward MN (1991) Causes and Predictability of Sahel Rainfall Variability. Climate Research Technical Note No. 12, Available from the National Meteorological Library, Meteorological Office, Bracknell, Berkshire, UK.

Semazzi FHM, Mehta V, Sud YC (1988) An investigation of the relationship between Sub-Saharan rainfall and global sea surface temperatures. Atmos-Ocean 26:118-138.

Shukla J, Fennessy MJ (1988) Prediction of time-mean atmospheric circulation and rainfall: Influence of Pacific sea surface temperature anomaly. J Atmos Sci 45:9-28.

Trenberth KE (1991) General characteristics of El Nino - Southern Oscillation. In: Glantz MH, Katz RW, Nicholls N (eds) Teleconnections linking worldwide climate anomalies, Cambridge University Press, Cambridge, p13.

Ward MN, Brooks S, Folland CK (1988) Predictability of seasonal rainfall in the northern Nordeste region of Brazil. In: Gregory S (Ed) Recent Climate Change, Belhaven, London, p237.

Ward MN, Folland CK (1991) Prediction of seasonal rainfall in the north Nordeste of Brazil using eigenvectors of sea-surface temperatures. Int J Clim 11:711-743.

Ward MN (1992) Provisionally corrected surface wind data, worldwide ocean-atmosphere surface fields and Sahel rainfall variability. J Climate, in press.

Young PC, Ng NC, Lane K, Parker DE (1991) Recursive Forecasting, Smoothing and Seasonal Adjustment of Non-Stationary Environmental Data. J Forecasting 10:57-90.

PREDICTABILITY OF SHORT-TERM CLIMATE VARIATIONS

J. Shukla
Center for Ocean-Land-Atmosphere Interactions
University of Maryland
College Park, MD 20742
USA

1. INTRODUCTION

It is well known that day-to-day changes of atmospheric motions are not predictable beyond a few weeks because of nonlinear dynamical interactions which can amplify even small initial errors due to incorrect observations used for defining the initial state of the atmosphere, or imperfect governing equations used for making predictions (Lorenz, 1965). However, is it possible that, despite the lack of predictability of day-to-day weather, monthly and seasonal averages of weather variable can be predicted with sufficient accuracy to be of some value for practical applications? In this paper, we attempt to answer this question by reviewing our current knowledge of observed variability and our understanding of physical processes that determine the interannual variability and predictability of monthly and seasonal averages. We also summarize the recent attempts to predict monthly and seasonal mean atmospheric circulation using dynamical models.

In this paper, the phrase 'short-term' refers to fluctuations with time scales varying from a month (~ 30 days) to a few years (~ 1000 days). Although our main emphasis will be on discussing the predictability of monthly (~ 0-30 day mean) and seasonal (0-90 day mean) variations from a given initial state, we also briefly address the question of predictability of the El Niño-Southern Oscillation (ENSO) phenomenon which has provided a basis for prediction of seasonal averages, several seasons in advance.

NATO ASI Series, Vol. I 6
Prediction of Interannual Climate Variations
Edited by J. Shukla

2. MECHANISMS FOR INTERANNUAL VARIABILITY

A convenient conceptual framework to describe and understand the mechanisms for interannual variability of monthly and seasonal average circulation is to consider the role of the internal dynamics and the boundary forcings separately (Shukla, 1981a). It must be recognized, however, that in reality the two are inseparable because the changes in the boundary forcings at the earth's surface (sea surface temperature (SST), soil wetness, snow cover and sea ice) are caused by their interactions with the overlying atmospheric motions.

Consider an idealized situation in which the boundary conditions of SST, soil wetness, snow cover and sea ice at the earth's surface, and the external forcings of the solar radiation are invariant with time, and a global climate model is integrated for several years. The simulated climate will undoubtedly exhibit interannual variability of monthly and seasonal mean circulation. These variations are caused by inherent dynamical instabilities, nonlinear interactions among different space and time scales and interactions of fluctuating zonal winds with orographic and thermal forcings. If observed variations in the real atmosphere were not significantly different from those produced by internal dynamical processes alone, the prospects of predicting the monthly and seasonal averages would not be much better than those for predicting the day-to-day weather fluctuations.

Based on a very large number of observational and modeling studies, it is now well established that changes in the boundary conditions of SST, soil wetness, snow cover and sea ice can produce large and systematic changes in atmospheric circulation. It is, therefore, natural to conclude that the predictability of monthly and seasonal average circulation could be significantly enhanced if the changes in these boundary conditions could also be predicted.

Fortunately, some of these boundary conditions change rather slowly compared to the changes in the atmospheric circulation and therefore, under certain situations, using realistic dynamical models the evolution and persistence of anomalous atmospheric circulation can be predicted with the assumption of persistence of the anomalous boundary conditions.

3. PREDICTABILITY OF MONTHLY AND SEASONAL MEAN ATMOSPHERIC CIRCULATION

For prediction of monthly and seasonal average circulation the initial condition of the atmosphere is important; but the boundary conditions at the earth's surface must also be correctly specified. The conceptual framework described here suggests that the predictability of time average atmospheric circulation should be different for the tropical and the extratropical flows (Shukla, 1981b). The tropical circulation consists of the large scale Hadley and Walker circulations whose fluctuations are determined mainly by changes in the boundary conditions at the earth's surface. The day-to-day weather changes in the tropics are associated with synoptic disturbances which, with the exception of tropical cyclones, are generally weaker compared to the large amplitude disturbances in mid-latitudes. The changes in the boundary conditions (viz. El Niño SST anomalies) can produce significant shifts in the locations of the ascending and descending branches of the Hadley and Walker circulations, giving rise to large shifts in seasonal mean rainfall patterns. In fact, the occurrence of seasonal tropical droughts and floods can be generally interpreted as spatial and temporal shifts in the climatological patterns of rainfall. The synoptic scale disturbances in the tropics are of secondary importance in affecting the location and intensity of large scale Hadley and Walker circulations.

The nature of variability in the mid-latitudes is distinctly different. The internal dynamical processes are the dominant mechanism for the day-to-day, as well as, the interannual variability. It is difficult to distinguish between the internal variability and the role of slowly varying boundary conditions. For example, a monthly mean circulation anomaly in the mid-latitudes can be very different depending upon the presence or the absence of a blocking event during that period. Thus, any success in predicting the monthly mean circulation anomaly will largely depend upon predicting the blocking event. Since there is no clear evidence that anomalous boundary conditions are the main forcing for the occurrence of blocking events, there is no guarantee that either a correct specification of the boundary conditions initially or even a correct prediction of the boundary conditions will help predict the blocking event and the corresponding monthly mean circulation.

The role of global boundary conditions in forcing mid-latitude variability remains an open question, in part, because GCM sensitivity experiments to address this question have not been carried out systematically. The available results are inconclusive. There have been several numerical simulations in which an atmospheric GCM is integrated with prescribed climatological SST and the variability is compared with the observed variability. If it is found that the two estimated variabilities are comparable - and they are for the extratropics - it is concluded that the variations of SST are not important because the model was able to simulate the observed variability even without accounting for SST variations. We would like to submit that this argument is flawed because the model simulated variability in the absence of the changes in the boundary conditions can be equal to the observed variability for a variety of other reasons depending upon the nature of global constraints imposed in the particular model.

3.1 Prediction of monthly and seasonal mean tropical circulation

We do not wish to dwell on this topic in any detail here because it has already been established, through a very large number of GCM sensitivity experiments, that the time mean tropical circulation and rainfall are highly predictable if the boundary conditions of SST and soil wetness are correctly specified (Shukla and Fennessy, 1988; Mo and Kalnay, 1991; WCRP, 1988).

The question of predictability of monthly averages about a seasonal average was examined by Fennessy and Shukla (1991) with correctly prescribed SST. It was found that there was little skill in predicting monthly departures from a seasonal mean. This generally has also been the experience for empirical predictions.

3.2 Prediction of monthly mean mid-latitude circulation (winter season)

Following the pioneering work of Miyakoda et al. (1983), large numbers of investigators have carried out hindcasts and forecasts for monthly mean atmospheric circulation using general circulation models (see the paper by Royer in this volume for a comprehensive review). In a study by Baumhefner (1991) in which he carried out an ensemble of forecasts for different initial states, he showed that there is considerable skill in prediction of monthly means. In these hindcast experiments, he used climatological boundary conditions of SST, soil moisture, snow cover and sea ice. It is likely that if he had used the observed boundary conditions, the skill of those hindcasts would have been even higher.

There also have been several classical predictability studies in which the initial state is randomly perturbed and the growth of initial error is examined for the monthly mean circulation. Using the NCAR community climate model with perpetual winter boundary conditions, Tribbia and Baumhefner (1988) found that the root mean square error between the twin model runs was comparable to the model variance for the first 30 day means, suggesting that the prospects for predicting thirty-day averages are not quite good. It is likely that this rather pessimistic result could be at least in part due to the model's deficiency in simulating the low frequency variability. We have further examined (Shukla, 1988) the predictability of monthly means using the two layer model by Schubert and Suarez (1989) and found that in contrast to the results of Tribbia and Baumhefner, for this model, 30-day averages for days 1-30, and even up to days 21-50, are highly predictable. We consider these overly optimistic results also suspect because the low frequency variability for this model was about 150% of the observed values where as it was about 67% for the NCAR model used by Tribbia and Baumhefner.

In an earlier paper (Shukla, 1981a), an analysis of variance was used to compare model integrations with large differences in the initial conditions, and with small random differences in the initial conditions. It was concluded that, even in the absence of suitable models to predict the boundary conditions at the earth's surface the atmospheric monthly means are dynamically predictable.

It is likely that a skillful prediction of monthly mean circulation might require better skill in predicting the 30-60 day oscillations (referred to as Madden-Julian oscillations). Krishnamurti (1991) has shown an example of skillful forecasts of up to a month by using observed SST and suppressing high frequency fluctuations.

Prediction of monthly means is a straightforward extension of numerical weather prediction. It is, however, quite important that, in addition to the usual initialization procedures to have an appropriate divergence field, the initial state must be completely consistent with the boundary conditions at the earth's surface (SST, soil moisture, snow, sea ice, etc.). Predictions beyond short range (0-5 days) should also include a description of different circulation regimes and estimates of ensemble variance.

Based on a large number of investigations made during the past ten years, it is reasonable to conclude that socially useful predictions of monthly mean atmospheric circulation can be produced routinely using dynamical models. However, it should be noted that especially for the extratropical regions, most of the skill in 30-day prediction comes from the first 15 days (e.g., Palmer et al., 1990).

3.3 Prediction of seasonal mean mid-latitude circulation (winter season)

There have only a few systematic investigations of predictability of seasonal mean circulation. If predictability of time averages were defined in terms of the ratio between the observed variances and natural variability (noise), as suggested by Madden (1976), seasonal averages are found to be more predictable than monthly averages (Singh, 1989). Twenty years of daily 500 mb analyses were utilized by Singh (1989) to calculate the ratio of the variance of observed seasonal means and noise in seasonal variance. The ratios were found to be higher than those for monthly means. This enhanced predictability is mainly an indication of the influence of the slowly varying boundary conditions. By taking a seasonal average, intra-monthly variations are filtered out and the effects of the boundary forcings can be detected. The notion of enhanced predictability of seasonal averages is further supported by model calculations of Lau (1985), for which Palmer (1987) showed that the ratio of observed variance to model simulated variance is far higher for the seasonal averages than that for the monthly averages.

Although studies of Shukla and Bangaru (1979), Palmer and Sun (1985), Lau and Nath (1990) and Wallace et al. (1990) show modeling and observational evidence of an association between mid-latitude SST and circulation, the possible influences of tropical and/or extratropical SST variations on the seasonal mean extratropical circulation are not clearly understood. One way, and perhaps the only way, to address this question is to make short term climate predictions with and without global SST variations starting from observed initial states and verify against actual observations. This is necessary because the mid-latitude response to tropical or extratropical SST changes depends on the structure and variability of the mid-latitude flow.

Recently, seasonal predictions (hindcasts) have been carried out by Brankovic et al. (1993) and Shukla and Marx (1993) using the ECMWF and the COLA models, respectively. Both studies have found significant effects of SST anomalies on seasonal predictability. In agreement with a large number of GCM sensitivity studies in the past, the impact of SST anomalies was especially large on the simulations of tropical and subtropical circulation.

3.4 Prediction of seasonal mean mid-latitude circulation (spring and summer season)

Following the conceptual model for the mechanisms for interannual variability, one may formulate the conjecture that, since the day-to-day fluctuations during the spring and summer season are rather weak, it should be possible to detect the influence of the boundary forcings even in the mid-latitudes.

Brankovic et al. (1993) have shown that in the spring season the skill of the ensemble mean seasonal forecast was highest and the spread of the ensemble was smallest among the four seasons.

It has been further shown by Fennessy et al. (1993) that changes in the initial specification of the soil wetness has significant effects on the simulated seasonal mean temperature and rainfall. The GCM simulation of 1988 drought over the U.S. was far more realistic when the initial soil wetness was based on the analysis-forecast cycle of the ECMWF model rather than climatology.

3.5 Prediction of seasonal averages several seasons in advance

The entire discussion in the preceding subsections refers to prediction of monthly (0-30 days) or seasonal (0-90 days) averages starting from a given initial state of the atmosphere and boundary conditions at the earth's surface. There is now a large body of modeling evidence that suggests that the coupled ocean-atmosphere system in the tropical Pacific is predictable up to several seasons (see Latif et al. (1993) for a review). The skill of these forecasts depends upon the initial state of the ocean (forecasts initiated in the summer and fall seasons are more skillful than those initiated in the winter season) suggesting that the initial state of the ocean contains the "memory" which makes it possible to predict the evolution of the coupled system for periods of one to two years. In a classical predictability study by Goswami and Shukla (1990) for the coupled model developed by Cane and Zebiak (1987), it has been shown that the doubling time for small perturbations in the initial state is a few months for the coupled system as compared to a few days for weather prediction models.

The existence of the extended range predictability of the coupled ocean-atmosphere system for the tropical Atlantic and the tropical Indian Oceans is yet to be shown using realistic models and realistic initial states of the ocean. It is likely that it would be more difficult to make skillful forecasts of SST anomalies in the tropical Atlantic and Indian Oceans because the interannual variability of SST is relatively small in these oceans compared to that in the tropical Pacific Ocean.

The prospects for prediction of extratropical SST several seasons in advance are not very good because of the strong dependence of SST on surface heat fluxes which are largely determined by the highly variable and unpredictable mid-latitude atmospheric variability.

4. SUMMARY AND CONCLUSIONS

- Although day-to-day weather fluctuations are not predictable beyond a few days, space-time averages of atmospheric circulation and rainfall are predictable for monthly and seasonal averages.
- Prediction of monthly means (0-30 days) is a straightforward extension of numerical weather prediction with one important difference: The definition of the initial state, in

addition to the usual initialization procedures to have an appropriate divergence field, now must have complete consistency between the atmosphere and the boundary conditions at the earth's surface (sea surface temperature, soil moisture, etc.).

- For prediction of seasonal means (0-90 days), the atmospheric initial condition remains important, but the boundary conditions at the earth's surface must also be predicted. (This has high potential for useful predictions in the tropics because SST changes slowly).
- If predictability of time averages is defined in terms of the ratio between observed variances and natural variability (noise), seasonal averages are more predictable than monthly averages, especially for the tropics.
- For prediction of seasonal means, for lead times of three to five seasons, the ocean initial condition is very important. The "memory" resides in the initial condition of the ocean and the coupling between the atmosphere and the ocean is the source of enhanced predictability of the coupled system.
- If tropical SST variations could be predicted at extended range (a few seasons to a few years), tropical rainfall and circulation could also be predicted with a high degree of confidence.
- The possible influences of tropical and/or extratropical SST variations on the extratropical circulation are not clearly understood.
- There is enough observational (statistical correlations) and modeling (sensitivity experiments) evidence to suggest that experimental prediction of seasonal means should be carried out routinely using the observed initial state of the atmosphere, observed global SST and soil wetness, a "good" GCM, and some scheme for prediction of the global SST.

REFERENCES

Baumhefner DP (1991) Evaluation of long-range forecast skill from Monte-Carlo ensemble integrations of a low-resolution NCAR Community Climate Model. Extended abstracts submitted to the ICTP/WMO international technical conference on long-range weather forecasting research (Trieste, Italy, 8-12 April 1991), WMO/TD n° 395; 15-19

Brankovic C, Palmer TN, Ferranti L (1993) Predictability of seasonal atmospheric variations. J Clim. To be submitted

Cane M, Zebiak SE (1987) Prediction of El Niño events using a physical model. In Atmospheric and Oceanic Variability. H Cattle, Ed., Royal Meteorological Society, 153-182

Fennessy MJ, Shukla J (1991) Comparison of the impact of the 1982/83 and 1986/87 Pacific SST anomalies on time-mean predictions of atmospheric circulation. J Clim 4: 407-423

Fennessy MJ, Kinter III LJ, Marx L, Sellers P, Shukla J (1993) Influence of initial soil wetness on GCM simulations of the 1988 US drought and heatwave. Proceedings of AMS Conference on Hydroclimatology, January 17-22, 1993, Anaheim CA

Goswami B, Shukla J (1991) Predictabillity of a coupled ocean-atmosphere model. J Clim 4: 3-22

Krishnamurti TN (1991) Simulation of low-frequency aspects of monsoons. 1990 ECMWF seminar on tropical of interactions

Latif M, Barnett TP, Cane MA, Flügel M, Graham NE, von Storch H, Xu J-S, Zebiak SE (1993) A review on ENSO prediction studies. Submitted to Climate Dynamics

Lau N-C, Nath MJ (1990) A general circulation model study of the atmospheric response to extratropical SST anomalies observed in 1950-79. J Clim 3: 965-989

Lau N-C (1985) Modelling the seasonal dependence of the atmospheric response to observed El Niños in 1962-1976. Mon Wea Rev 113:, 1970-1996

Lorenz EN (1965) A study of the predictability of a 28-variable atmospheric model. Tellus 34: 505-51

Madden RA (1976) Estimates of the natural variability of time-averaged sea level pressure. Mon Wea Rev 104: 1942-952

Miyakoda K, Gordon CT, Caverly R, Stern WF, Sirutis J, Burke W (1983) Simulation of a blocking event in January 1977. Mon Wea Rev 111: 846-869

Mo KC, Kalnay E (1991) Impact of sea surface anomalies on the skill of monthly forecasts. Mon Wea Rev 119: 2771-2793

Palmer TN (1987) Modelling atmospheric low-frequency variability. In: Atmospheric and Oceanic Variability, Editor: H. Cattle, Royal Meteorological Society

Palmer TN, Sun Z (1985) A modelling and observational study of the relationship between sea surface temperature anomalies in the north-west Atlantic and the atmospheric general circulation. Q J Roy Met Soc 111: 947-975

Palmer TN, Brankovic C, Molteni F, Tibaldi S, Ferranti L, Hollingsworth A, Cubasch U, Klinker E (1990) The European Center for Medium-Range Weather Forecasts (ECMWF) Program on extended-range prfediction. Bull Amer Meteorol Soc 71: 1317-1330

Royer J-F (1991) Review of recent advances in dynamical extended-range forecasting for the extratropics. Proceedings of the NATO workshop in *Prediction of Interannual Climate Variations*. NATO ASI Series Publication. **Global Environmental Change.** Springer-Verlag

Schubert SD, Suarez M (1989) Dynamical predictability in a simple general circulation model: Average error growth. J Atmos Sci 46: 353-370

Shukla J (1981a) Dynamical predictability of monthly means. J Atmos Sci 38: 2547-2572

Shukla J (1981b) Predictability of the tropical atmosphere. NASA Tech Memo 83829: pp 51

Shukla J (1984) Predictability of time averages, **Part II:** The influence of the boundary forcing. In *Problems and prospects in long and medium range weather forecasting*, D. M. Burridge and E. Kallen, Eds. Springer-Verlag New York, 155-206

Shukla J (1988) Predictability of monthly mean circulation and rainfall: **Part I:** Classical dynamical predictability experiments; and **Part II:** Influences of 1982-83 and 1986-87 El Niño SST anomalies. Proceeedings of the Workshop on Predictability in the Medium and Extended Range (16-18 May 1988), ECMWF, England, 251-261

Shukla J, Bangaru B (1979) Effect of a Pacific sea-surface temperature anomaly on the circulation over North America: A numerical experiment with the GLAS model. GARP Publication Series 22: 501-518

Shukla J, Fennessy MJ (1988) Prediction of time-mean atmospheric circulation and rainfall: Influence of Pacific sea surface temperature anomaly. J Atmos Sci 45: 9-28

Shukla J, Marx L (1993) Prediction of seasonal mean circulation. To be submitted to Mon Wea Rev

Singh RR (1989) Potential predictability of 500 mb geophotential heights on both monthly and seasonal scales over the northern hemisphere. MS thesis, University of Maryland, College Park, Maryland

Tribbia JJ, Baumhefner DP (1988) Estimates of the predictability of low-frequency variability with a spectral general circulatin model. J Atmos Sci 16: 2306-2317

Wallace JM, Smith, Jiang (1990) Special Patterns of atmosphere-ocean interaction in the Northern Winter. J Clim 3: 990-998

Walsh JE, Ross B (1988) Sensitivity of 30-day dyanmical forecasts to continental snow cover. J Climate 1: 739-754

WCRP (1988) Modelling the sensitivity and variations of the ocean-atmosphere system. WCRP-15. WMO Geneva

PREDICTABILITY OF CLIMATE

Antonio Speranza
University of Camerino
Department of Physics
Via Madonna della Carceri
62032 Camerino (MC)
ITALY

1. INTRODUCTION

The theoretical foundations of the problem of deterministic predictability are nowadays quite clear. Let us summarize the basic concept and definitions.

The spectrum of Liapunov exponents provides a summary of the local stability properties along a nonlinear trajectory in the phase-space of a dynamical system.

The stability properties of a system are determined by its behaviour under small perturbations. All possible perturbations can be examined simultaneously by following the evolution of an ensemble of points initially contained in a small N-dimensional ball, where N is the phase space dimension. In practice, if the nonlinear trajectory is given by:

$$X = A\ (X), \qquad (1)$$

where X is the N-dimensional vector in the phase space and A (X) the nonlinear operator, the time evolution of infinitesimal perturbations along the trajectory can be computed by simultaneously integrating the linear equations:

$$X' = [J(X(t))]\ X' \qquad (2)$$

where [J] is the N x N Jacobian matrix computed along the trajectory, and where X' denotes a small deviation from the trajectory itself. In cases in which the Jacobian matrix can be analytically defined and computed along the trajectory a standard technique is usually used to compute the exponents. Otherwise different approaches must be devised.

The computation of the first Liapunov exponent proceeds as follows: chosen an initial X'(o), and integrated eq. (2) along the orbit X(t), the quantity d(t) = || is obtained, where, for convenience, the initial norm d(0) is chosen to be unity. The practical difficulty is that the

NATO ASI Series, Vol. I 6
Prediction of Interannual Climate Variations
Edited by J. Shukla

norm of X' increases exponentially in time, leading to overflow. To circumvent this problem, the usual approach consists on choosing a small fixed time interval τ, and renormalizing X' to a norm of unity every τ time units. Thus, one iteratively computes:

$$\begin{aligned} d_k &= \| X_{k-1}(\tau) \| \\ X_k(0) &= X_{k-1}(\tau)/d_k \end{aligned} \tag{3}$$

From (3) it is possible to define the quantity:

$$\Sigma_n^{(1)} = 1/(n\tau) \sum_{t=1}^{n} \ln (d_t) \; .$$

For τ not too large, it can be shown that

$$\sigma^{(1)} = \lim_{n \to\!> \infty} \sigma^{(1)}$$

exists and is independent of τ (see Lichtenberg and Lieberman, (1983). It is possible now to calculate the entire set of Liapunov exponents $\sigma^{(1)}$, $\sigma^{(2)}$, etc., by choosing initial vectors $X'^{(1)}$, $X'^{(2)}$, etc., along orthogonal directions must fail due to numerical errors, which eventually bring $X'^{(p)}$ into alignment with $X'^{(1)}$. To circumvent this problem, $\{X'^{(p)}\}$ must be orthonormalized every τ time unites by means of a Gram-Schmidt procedure. Thus, one can calculate for each time interval τ the quantities:

$$u_{k-1}^{(p)} (\tau) = X'^{(p)}_{k-1} - \sum_{i-1}^{p-1} (X'^{(i)}k \; (0) \cdot X'^{(p)}_{k-1} (\tau)) \cdot X'^{(i)}_{k}(0)$$

$$d_k^{(p)} = | \; u_{k-1}^{(p)} (\tau) \; |$$

$$X'^{(p)}_{k} (0) = u_{k-1}^{(p)} (\tau)/d_k^{(p)}, \quad p \geq 2$$

from which to obtain the p-th Liapunov exponent:

$$\sigma^{(p)} = \lim_{n \to\to \infty} 1/(n\tau) \sum_{i=1}^{n} \ln (d_i^{(p)}) \ . \tag{4}$$

Given the spectrum of Liapunov exponents, an estimate of the information dimension and of metric entropy can be given. An upper bound to the information dimension (Kaplan and Yorke, 1979) is given by:

$$\sigma \leq j + \sum_{p=1}^{J} \sigma^{(p)} \ / \ | \ \sigma^{J+1} \ | \tag{5}$$

where J is the largest integer for which

$$\sum_{p=1}^{J} \sigma^{(p)} >> 0 \ .$$

Qualitatively, this definition can be understood by noticing that the sum of the first q exponents is the divergence of a q-dimensional volume in phase space.

Liapunov exponents can be used to estimate predictability from the average divergence rate of an infinitesimally small phase space volume:

$$K \leq \sum_{p} \sigma^{(p)} \ , \ \sigma^{(p)} \geq 0 \ .$$

where K is known as metric (Kolmogorov) entropy (see Farmer, 1982), and represents the generation rate of new information. The inequalities (5) and (6) are known to be saturated if the natural measure is completely continuous in all directions with positive $\sigma^{(p)}$.

It is also possible to estimate lower bounds for both σ and K. The order 2 generalized entropy, K_2, and the correlation dimension v are often considered as convenient lower bounds for K and σ (see Grassberger and Procaccia, 1983, 1984; Fraedrich, 1986).

There is a general feeling in the literature that the computation of K, v and other estimates of statistical properties of dynamical systems requires, in the case of large dimensionality, extremely long time series (see, for example, Ruelle, 1990). Despite of the caveats expressed by mathematicians several researchers have produced estimates of the above coefficients based on the analysis of observational records of atmospheric motion. I believe we are still very far from achieving correct estimates of such dimensions from observations even for low dimensions of the attractor, because of the lack of independent observations. For example, the decorrelation time of "weather" in tropospheric records is about 5 days. This reduces the number of independent observations to a few tens in a year and this make it impossible to work on more than a few thousands of data: unless the dimension of the "tropospheric attractor is extremely small we are out of business! Ever more restrictive conditions apply to paleoclimate.

In order to learn something it is here useful to make use of numerical models sufficiently complex to display "almost turbulent" behaviour, but sufficiently simple to be integrated for very long times. This puts us in the range of hundreds of components. The specific example was investigated by Speranza-Malguzzi (SM from now on) and is discussed in the following section.

2. AN INTERMEDIATE DIMENSIONALITY MODEL

The model used in the numerical experiments discussed below is deduced from the two-layer, quasigeostrophic equations:

$$a_t \nabla^2 \phi + J(\phi, \nabla^2\phi + \beta y) + J(\tau, \nabla^2\tau) = -\frac{VE}{2} \nabla^2(\phi - \tau) \tag{7}$$

$$a_t (\nabla^2\tau - 2F\tau) + J(\tau, \nabla^2\phi - \beta y)\, J(\phi, \nabla^2\tau) - 2FJ(\phi, \tau) =$$
$$2FJ(\phi, \tau) = \frac{v_E}{2} \nabla^2(\phi - \tau) - vs\, \nabla^2\tau + F v_H (\tau - \tau*). \tag{8}$$

Here $\phi = (\psi_1 + \psi_3)/2$ and $\tau = (\psi_1 - \psi_3)/2$ are, respectively, the barotropic and the baroclinic (temperature) components of the streamfunction $\psi_{1,3}$; J, the Jacobian operator; ∇^2 , the Laplacian operator; F, the Froude number; v_E , $_v S$, and v_H are, respectively, the coefficients of Ekman pumping at the lower surface, friction at the surface separating the two layers, and Newtonian cooling; β, the gradient of the Coriolis parameter; and $\tau*$, the source which forces the fluid motion (see, for instance, Lorenz, 1963). The variables are in the standard nondimensional form of Pedlosky (1979).

We separate the zonally symmetric part of the vertically averaged velocity and vertical shear, U(y,t) and m(y,5) respectively:

$$\phi = -\int U\ (y,\ t)\ dy + \phi'\ (x,\ y,\ t) \tag{9}$$

$$\tau = -\int m\ (y,\ t)\ dy + \tau'\ (x,\ y,\ t). \tag{10}$$

We introduce a particular form of the solution for the nonsymmetric component:

$$\phi' = \sum_{n=1} A_n\ (y,\ t)\ \exp\ (inkx) + (*) \tag{11}$$

$$\tau' = \sum_{n=1} B_n\ (y,\ t)\ \exp\ (inks) + (*) \tag{12}$$

where (*) indicates the complex conjugate of the proceeding term. Here we truncate to just one Fourier component:

$$\{\exp\ \{inkx\},\ n = 1\}\ . \tag{13}$$

As shown by SM, this is sufficient to produce an "essential" dynamics of maintenance of the middle latitude jet. As discussed in the introduction, by "essential" we mean that the degrees of freedom of the model must show, for a realistic range of the external parameters, chaotic behaviour with unimodal probability density distribution. With assumptions (13), eq. (9) and (10) reduce to:

$$\phi + -\int U\ dy + A\ \exp\ (ikx) + (*)$$
$$\tau = -\int m\ dy + B\ \exp\ (ikx) + (*)$$

and by projecting (7) and (8) onto exp (ikx) we obtain:

$$U + 1/2\ \nu_E\ (U-m) + 2\ K \cdot Im\ \{AA_{yy}* + BB_{yy}*\} = 0 \tag{14}$$

where

$$m_{yy} - 2\,F\,m + v_s\,m_{yy} - 1/2\,v_E\,(U-m)_{yy} - 2\,F\,v_H\,(m-m*) + 4\,F\cdot Im\,\{A*\,B\}_{yy} + 2\,k\cdot Im\,\{A\,B_y* + B\,A_y*\}_{yyy} = 0 \tag{15}$$

$$A_{yy}\,K^2\,A + (1/2\,v_E + [kU)\cdot(k^2\,m + m_{yy} - \beta) + 1/2\,v_E\,k^2]\cdot B + (ikm - 1/2\,v_E)\cdot B_{yy} = 0 \tag{16}$$

$$b_{yy} - k^2 - 2\,F\,B + (1/2\,v_E + v_s + iku)\cdot B_{yy} - [ik\cdot(k^2\,U + U_{yy} - \beta)\; 1/2\,v_E\,k^2]\cdot B - (V_S\,K^2 + 2\,F\,v_H + 2\,ik\,F\,U)\cdot B + (ikm - 1/2\,v_E)\cdot Ayy - [ik\cdot(k^2\,m + m_{yy}) - 1/2\,v_E\,k^2 - 2\,ikFm]\cdot A = 0 \tag{17}$$

$$\tau * = - \int m * (y)\, dy\,.$$

and where, because of the simple wavelike form of the nonsymmetric disturbances, all the wave-wave interaction coefficients vanish and the only nonlinear interactions take place between symmetric and nonsymmetric components (latitudinal convergence of eddy momentum flux and heat flux). Eqs. (14)-(17) constitute a set of six real field equations in latitude y and time t, with coefficients A and B being complex. As lateral boundary conditions we assume rigid walls at y = 0, L_y, which imply U = m = A = B = O at the boundaries. System (8)-(11) can be transformed into a set of ordinary differential equations by further decomposing U, m, A, B in Fourier series:

$$\begin{pmatrix} U \\ m \\ A \\ B \end{pmatrix} = \sum_{J=1}^{JT} \begin{pmatrix} U_j \\ m_j \\ A_j \\ B_j \end{pmatrix} \sin\,(j\,\pi\,y/L_y)$$

where the total number of Fourier components is denoted by JT. The system is driven in eq. (15) through the term 2 F v_H (m - m*) which forces a relaxation to the radiative equilibrium potential temperature profile will be assumed of the form:

$$\theta * = 2\,\tau * = \theta_E\,\cos\,(\pi y/L_y)$$

which corresponds to the vertical shear

The parameter θ_E (representing half of the equator to pole temperature contrast at the

TABLE 1

Spectrum of Liapunov exponents $\sigma^{(p)}$, in decreasing order, for JT16 (p=1.50) (a); and JT32 (p=1.80) (b). Units are dimensionless.

0.565	0.488	0.422	0.373	0.323	0.285
0.248	0.220	0.187	0.157	0.138	0.118
0.096	0.083	0.062	0.050	0.037	0.022
0.010	0.001	− 0.007	− 0.019	− 0.028	− 0.039
− 0.049	− 0.060	− 0.070	− 0.082	− 0.088	− 0.098 *a*
− 0.105	− 0.113	− 0.126	− 0.135	− 0.147	− 0.159
− 0.170	− 0.184	− 0.190	− 0.208	− 0.219	− 0.233
− 0.242	− 0.256	− 0.272	− 0.282	− 0.299	− 0.309
− 0.326	− 0.340				
0.469	0.427	0.373	0.329	0.295	0.263
0.241	0.210	0.190	0.173	0.155	0.132
0.123	0.114	0.098	0.088	0.076	0.067
0.055	0.048	0.046	0.038	0.025	0.022
0.015	0.010	0.006	0.004	− 0.002	− 0.007
− 0.014	− 0.015	− 0.021	− 0.022	− 0.028	− 0.030
− 0.035	− 0.038	− 0.041	− 0.043	− 0.050	− 0.051 *b*
− 0.053	− 0.057	− 0.059	− 0.062	− 0.065	− 0.067
− 0.071	− 0.071	− 0.077	− 0.076	− 0.080	− 0.082
− 0.087	− 0.089	− 0.092	− 0.094	− 0.095	− 0.101
− 0.103	− 0.104	− 0.109	− 0.111	− 0.116	− 0.116
− 0.124	− 0.125	− 0.127	− 0.133	− 0.136	− 0.137
− 0.141	− 0.147	− 0.148	− 0.156	− 0.159	− 0.161
− 0.167	− 0.175				

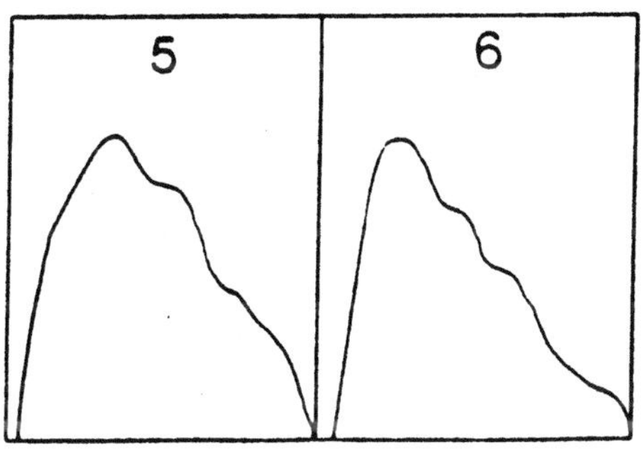

$$m * = \theta_E \, \pi/(2 \, l_y) \sin(\pi y/L_y) \; .$$

radiative equilibrium) will be determined empirically in such a way as to give the desired behavior of our model atmosphere.

System (14)-(17) describes the nonlinear equalibration of a single baroclinic wave through interactions with the zonal component. Statistical properties of (14)-(17) have been studied in SM; the state variables are characterized by a unimodal density distribution appropriate for the baroclinic medium-short wave components in the middle latitude circulation.

We use the techniques described above to compute the spectrum of Liapunov exponents. The parameter values used in the following can be found in SM and where chosen in such a way as to reproduce some aspects of the statistics of the mid-latitude jet. In particular, the parameter θ_E is set to $\theta_E = 9$. We integrate the truncated version of eqs. (14)-(17) starting from an arbitrary initial condition, and discarding the first two month of integration.

We compute the spectrum of Liapunov exponents for two cases with different truncations: JT = 16 and JT - 32 (hereafter referred to as JT16 and JT32). The value of the time interval π (see Section 3) is chosen empirically to be $\tau = 5$ (one time unit is 10^5s). The results presented below do not depend significantly upon the value of τ; for the case JT16 it was found that, by varying τ in the range from 0.5 to 10, the uncertainty in the estimate of the exponents is the order of a few percents.

Because of computer costs, JT16 was run for 640 x τ time units (10 years), and JT32 for 400 x τ time units only. Given the particular value of θ , the two cases show similar statistical properties, in contrast to the case JT = 8 (not shown). As an example, Figures 1a,b show, respectively, the scatter diagram in the (U_1, m_1) plane for JT16 and JT32. Table 1 shows the spectrum of Liapunov exponents for JT16 and JT32, respectively. The convergence of the estimate of the spectrum stems from inspection of Figure 2 which shows, for JT32, the first 20 estimates of Liapunov exponents as a function of the number of iterations n (see expression (4)). Computation were made on a CRAY-XMP computer with single precision.

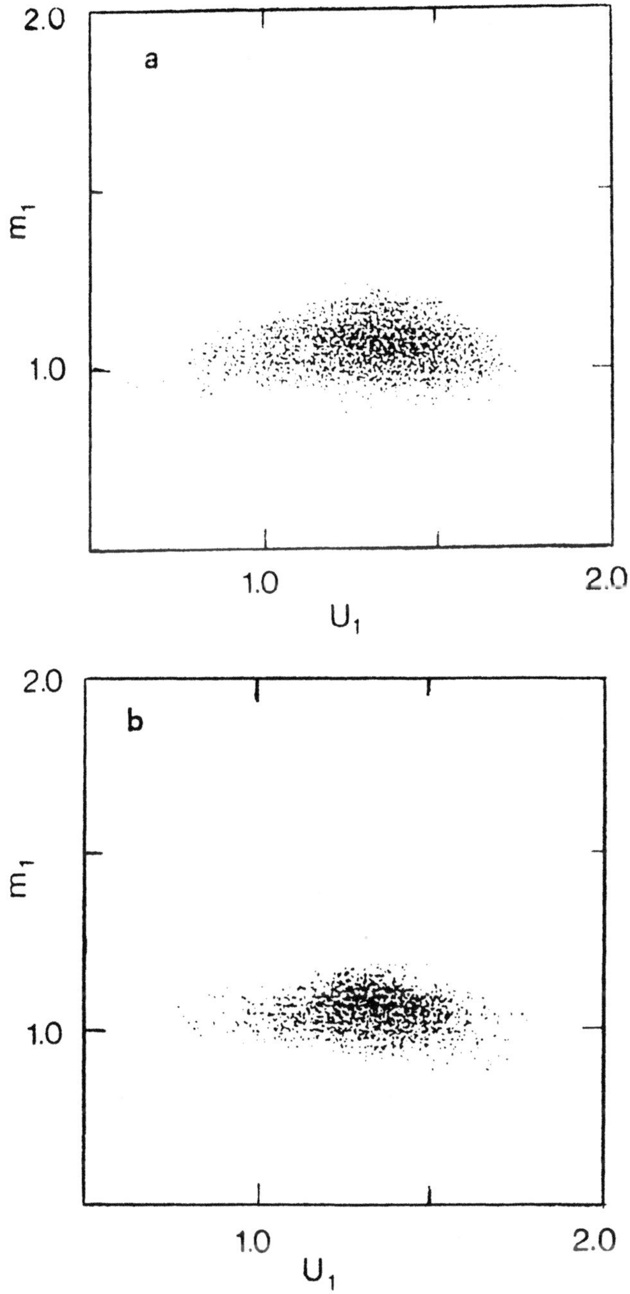

FIGURE 1: Scatter diagram in the (U_1, m_1) plane for JT16 (**a**) and JT32 (**b**).

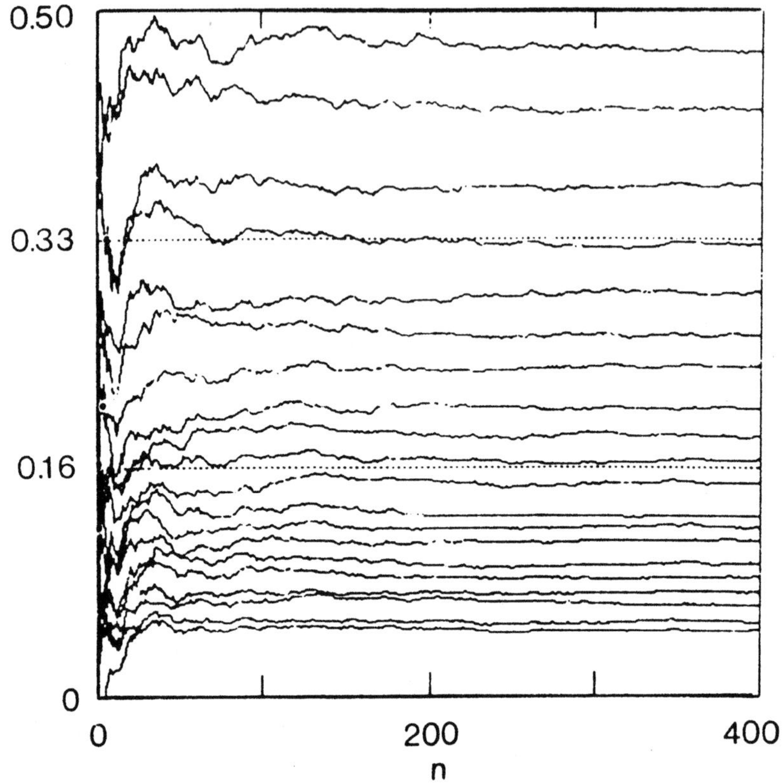

FIGURE 2: Estimates of the first 20 Liapunov exponents for JT32 as a function of the number n of iterations.

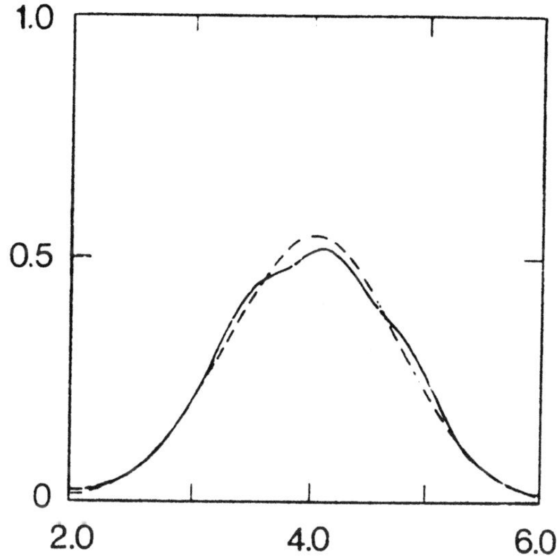

FIGURE 3: MPL estimate of the probability density of the "instantaneous metric entropy" for JT16 (continuous line). The dashed line is Gaussian having the same mean and variance.

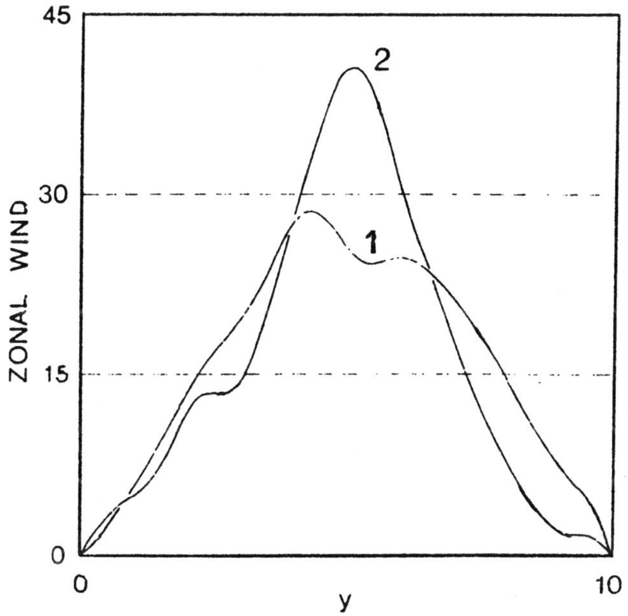

FIGURE 4: Composite of upper level zonal wind, as a function of latitude, over all cases with entropy ≤ 3 (curve labelled 1), and ≥ 5 (curve labelled 2).

It can be easily seen that the number of positive exponents, and their numerical values, are quite different in the two cases: in particular, JT16 shows the Liapunov dimension to be about 47 while for JT32, it is about 77.

However, it is noteworthy that the sum of positive exponents is approximately the same in the two cases; 3.9 for JT16 and 4.1 for JT32. Since the number of positive exponents increases with increasing resolution, a constant sum implies the lowering of their numerical values. A similar behaviour is found in other infinite dimensional dynamical systems, such as the delay equation (Farmer, 1982).

A particular Liapunov exponent is a global property of the attractor, as it is the time average, over an orbit virtually visiting the attractor, of the "instantaneous" mean over τ time units) growth or decay of a perturbation along a certain direction. Hence, we can associate a probability distribution function to each exponent. In the same way, we can define an "instantaneous metric entropy" by adding over all the instantaneous positive exponents. In this way, we can distinguish between more or less predictable situations, in terms of information generation rate along the trajectory. Figure 3 shows an estimate of the probability density function of such an "instantaneous metric entropy" for JT16. States more or less predictable than the average belong, respectively, to the left or right tail of the distribution. These two composites of states correspond to different synoptic situation, as can be seen from an inspection of Figure 4: the first composite is over all cases with "metric entropy" less than 3 while the second one corresponds to "metric entropies" greater than 5. The asymmetries of these profiles around the center of the domain are due to the small number of cases considered: 41 cases out of 640 fall below the lower cut-off threshold, while 70 cases are found above the upper one. On average, a less pronounced jet corresponds to a more predictable situation, while the opposite is true for a sharper jet. It is interesting to notice that the linear instability analysis of these two zonal winds gives a maximum growth rate of 0.29 for the more predictable case and 0.13 for the less predictable one, showing again the inadequacy of linear analysis of averaged states to assess predictability. This result confirms the difficulty, underlined in SM, in deriving meaningful dynamical knowledge from stability analysis of averaged states.

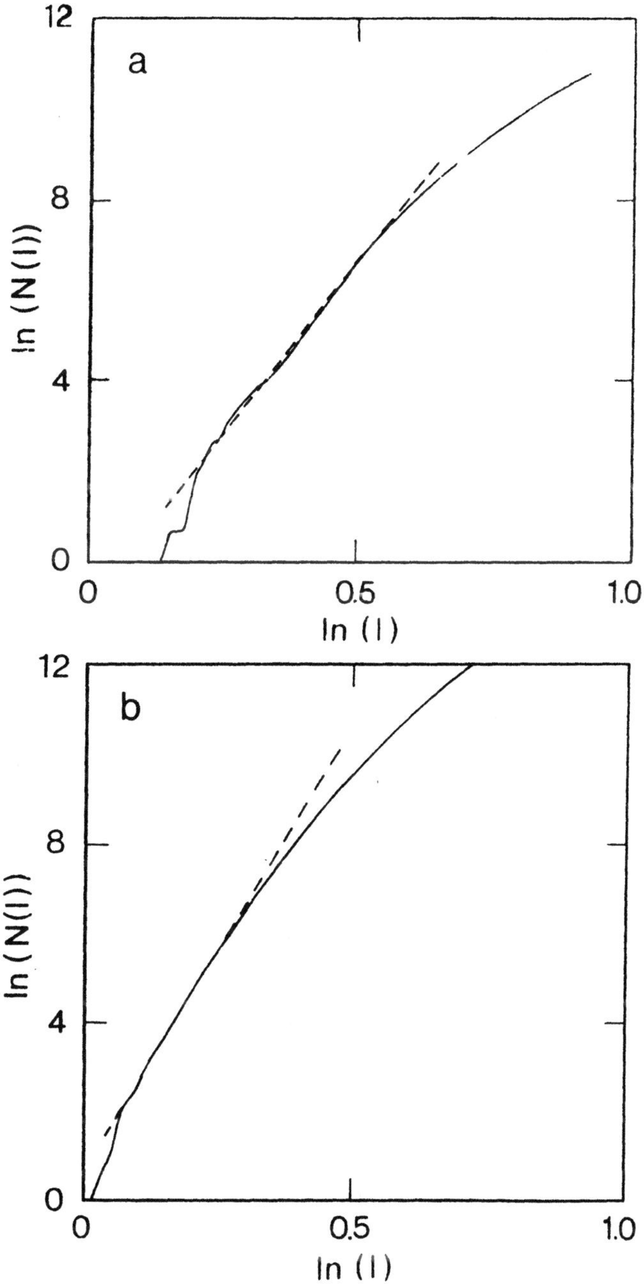

FIGURE 5: Logarithm of N (l) versus logarithm of distance l for JT16 obtained with 640 points (a) and 3000 points (b). The dashed line is the tangent to the curve at small distances.

We have found upper bounds to the metric entropy and information dimension from the estimate of Liapunov exponents. Lower bounds for the same quantities are provided by K_2 and v, respectively. The correlation dimension is usually computed from the distribution of the distance between pairs of points on a single variable trajectory evolving in a space which embeds the attractor. The procedure turns out particularly feasible when dealing with real data (see Guckenheimer and Buzyna, 1983, for the transition to turbulence in a rotating annulus; Nicolis and Nicolis, 1984, and Fraedrich, 1986, for the weather and climate attractors). Denoting the number of pairs of point that are up to a distance l part by N (l), the correlation dimension is comported from the relationship

$$N\ (1) \approx 1^{\nu} \tag{18}$$

where the total number of points goes to infinity and where l is small. Figure 5a shows ln (N(l)) versus ln (l) for JT16. For this experiment the number of points on the attractor at our disposal is 640,; thus, we can compute 640 x (640 - 1)/2 distances. A linear relationship between ln (N(l)) and ln (l) can be observed for a limited range of distances, which would indicate a correlation dimension of about 15 (for small distances (18) breaks down because there are not pairs of point, while for large distances there is obviously a saturation). Due to the high dimensionality found, it is doubtful that the number of points used in this computation is adequately large. We thus perform an additional integration of the model with JT = 16, thus obtaining a total of 3000 points on the attractor. Figure 5b shows that the new estimate of is about 20. Clearly, it is not possible to say anything about convergence of the estimate, since the slope for small distances is still increasing. However, if we assume that the slope increases monotonously with the number of points, we may say that $v \geq 19$ and, thus, that $19 \leq \sigma \leq 47$. It is unfeasible to attempt the computation of K_2 for such a large system given the present resources.

3. CONCLUSIONS

Models like the one here discussed have also been used for determining the properties of error growth (Trevisan, Malguzzi, Fantini, 1991), ultralong wave propagation in a baroclinic jet (Speranza and Malguzzi, 1982; Benzi and Malguzzi, 1991), etc. The results seem to support the idea that the dimension of the attractor of systems as complex as the tropospheric jet may be quite high, in effect, so high as to make it not feasible an estimate based on observations. Consequently, the only possible approach is through numerical simulation.

REFERENCES

Benzi R, Malguzzi P, Speranza A and Sutera A (1986) The statistical properties of general atmospheric circulation: Observational evidence and a minimal theory of bimodality. Q J Roy Meteorol Soc 112: 661-674

Benzi R, Sutera A and Jona-Lasinio G (1988) Stochastically perturbede Landau-Ginzburg equations. (submitted to J Stat Phys)

Buzzi A and Tosi E (1988) Low frequency variability of the atmospheric circulation: A comparison of statistical properties in both hemispheres and extreme seasons. Nuovo Cimento C 11: 467-488

Charney JG and DeVore JG (1979) Multiple flow equilibria in the atmosphere and blocking. J Atmos Sci 36: 1205-1216

Dalcher A and Kalnay E (1987) Error growth and predictability in operational ECMWF forecasts. Tellus 39A: 474-491

Defant A (1921) Die zerkulation der atmosphäre in den gemassigten breiten der erte. Geograf Ann 3: 209-266

Douglas CKM (1931) A problem of the general circulation. Q J Roy Meteor Soc 57: 423-431

Eckmann JP and Procaccia L (1986) Fluctuations of dynamical scaling indices in nonlinear systems. Phys Rev A 34: 659-661

Farmer JD (1982) Chaotic attractors of an infinite-dimensional dynamical system. Physica 4D: 336-393

Feigenbaum MJ, Kadanoff LP and Shenker SJ (1982) Quasi-periodicity in dissipative systems: A renormalization group analysis. Physica 3D: 370-386

Fraedrich K (1986) Estimating the dimensions of weather and climate attractors. J Atmos Sci 43: 419-432

Frederiksen JS (1978) Growth rates and phase speeds of baroclinic waves in multi-level models on a shpere. J Atmos Sci 35: 1816-1826

Grassberger P and Procaccia I (1983) Characterization of strange attractors. Phy Rev Lett 50: 448-451

Grassberger P and Procaccia I (1984) Dimensions and entropies of strange attractors from a fluctuating dynamical approach. Physica 13D: 34-54

Guckenheimer J and Buzyna G (1983) Dimension measurements for geostrophic turbulence. Phys Rev. Lett 51: 1438-1441

Hart JE (1979) Finite amplitude baroclinic instability. Ann Rev Fluid Mech 11: 147-172

Hoskins BJ (1983) Large-scale dynamic processes in the atmosphere. Academic Press

Hussain, AKMF (1983) Coherent structures-reality and myth. Phys Fluids 26(10): 2816-2850

Jeffreys H (1926) On the dynamics of geostrophic winds. Q J Roy Meteor Soc 52: 85-104

Kaplan J and Yorke J (1979) Functional differential equations and approximation of fixed points. In: HO Peitigen and HO Walther Speringer-Verlag 228

Kubicek M and Marek M (1983) Computational methods in bifurcation theory and dissipative structures. Springer-Verlag

Lichtenberg AJ and Lieberman MA (1983) Regular and stochastic motion. Springer-Verlag 259-285

Lorenz EN (1963) The mechanics of vacillation. J Atmos Sci 20: 448-464

Lorenz EN (1967) The nature and theory of the general circulation of the atmosphere. WMO

Lorenz EN (1969) Atmospheric predictability as revealed by naturally occurring analogues. J Atmos Sci 26: 363-646

Lorenz EN (1980) Attractors sets and quasi-geostrophic equilibrium. J Atmos Sci 37: 1685-1699

Lorenz EN (1982) Atmospheric predictability experiments with a large numerical model. Tellus 34: 505-513

Malguzzi P, Trevisan A and Speranza A (1990) Statistics and predictability for an intermediate dimensionality model of the baroclinic jet. Ann Geoph 8: 29-36

Nicolis C and Nicolis G (1984) Is there a climatic attractor? Nature 311: 529-532

Pedlosky J (1970) Finite amplitude baroclinic waves. J Atmos Sci 27: 15-30

Pedlosky J (1979) Geophysical fluid dynamics. Springer-Verlag New York 315-338

Saltzman B (1968) Steady state solutions for axially symmetric climatic variables. Pure Appl Geophys 69: 237-259

Saltzman B (1978) A survey of statistical-dynamical models of the terrestrial climate. Adv Geophys 20: 183-304

Saltzman B and Vernekar AD (1968) A parameterization of the large scale eddy flux of relative angular momentum. Mon Wea Rev 96: 854-857

Saltzman B and Vernekar AD (1971) An equilibrium solution for the axially symmetric component of the Earth's macroclimate. J Geophys Res 76: 1498-1524

Schubert SD and Suarez M (1989) Dynamical predictability in a simple General Circulation Model: Average error growth. J Atmos Sci 46: 353-370

Simmons AJ (1982) the forcing of stationary wave motion by tropical diabatic heating. Quart J Roy Meteor Soc 108: 503-534

Simmons AJ and Hoskins BJ (1978) The life cycles of some nonlinear baroclinic waves. J Atmos Sci 35: 414-432

Sparrow C (1982) The Lorenz Equations: Bifurcations, Chaos, and Strange Attractors. Springer-Verlag

Speranza A and Malguzzi P (1986) The response of atmospheric circulation to anomalous tropical heating: A re-examination of the theory of teleconnections in the context of turbulence theory. Study week: "Persistent Meteo-Oceanographic Anomalies and Teleconnections" Rome

Speranz A and Malguzzi P (1988) The statistical properties of a zonal jet in a baroclinic atmosphere: A semilinear approach. Part I: Quasigeostrophic, two-layer model atmosphere. J Atmos Sci 45: 3046-3061

Trevisan A, Malguzzi P and Fantini M (1990) A note on Lorenz's law for the growth of large and small errors in the atmosphere. J Atmos Sci (in print)

ON THE PROBLEM OF PREDICTION BEYOND THE DETERMINISTIC RANGE

J. Tribbia and D. Baumhefner
National Center for Atmospheric Research*
1850 Table Mesa Drive
Boulder, CO 80307-3000 USA

INTRODUCTION

Over the past two decades global numerical models of the atmosphere have become the primary tool used in the prediction of the weather out to ten days in advance. The success of these models as witnessed by the improvement in forecast skill in that time has encouraged researchers to investigate the utility and accuracy of forecasts of lower frequency atmospheric variability as obtained through the integration of these models beyond the limits of deterministic predictability. The problem of forecasting beyond this limit requires that the very idea of prediction be reformulated and viewed in a probabilistic sense in order to take into account the inherent loss of determinism of the physical system. In order to rationally forecast, one must redefine the goals of prediction in both a physical and mathematical sense and thereby limit the influence of the loss of determinism through the modifications of the definition of the prediction problem.

In order to proceed further it is thus necessary to first examine the general prediction problem on various time scales from a phenomenological perspective and then to delineate the goals of prediction at each time range, beginning at short-to-medium range deterministic prediction and ending at seasonal and interannual non-deterministic forecasting. It should be noted that the perspective taken will be strongly slanted towards the problems of prediction in middle and high latitudes, although some attempt will be made to note exceptions as they might exist in the tropics.

Prediction on various time scales

The problem of mid-latitude weather prediction is conveniently broken up both phenomenologically and conceptually into distinct time ranges. The most familiar predictions are the short-to-medium range forecasts of the weather in detail. For

*The National Center for Atmospheric Research is sponsored by the National Science Foundation.

NATO ASI Series, Vol. I 6
Prediction of Interannual Climate Variations
Edited by J. Shukla

predictions beyond 24 hours, the weather phenomenon to be simulated and predicted is the traveling synoptic scale system with its concommitant meteorological elements of precipitation. As is well known from a number of studies, the prediction of the temporal evolution of such features is highly sensitive to variations in the initial state and to small errors in the modelled time tendencies (Lorenz, 1969; Thompson, 1957; Leith, 1971; Smagorinsky, 1969; and recently, Shukla, 1981; Tribbia and Baumhefner, 1988 (TB)). This sensitivity is reflected in the fact that small errors amplify and eventually render forecasts useless as demonstrated in Fig. 1 from TB. Viewed conceptually from a dynamical systems perspective, deterministic predictions of this sort require that the trajectory of the numerical prediction match identically that of the atmosphere within the model phase space. The upshot of the predictability studies cited above is that small errors in the trajectory double in amplitude in approximately two days so that accurate deterministic predictions beyond two weeks are nearly impossible (cf. Figure 1).

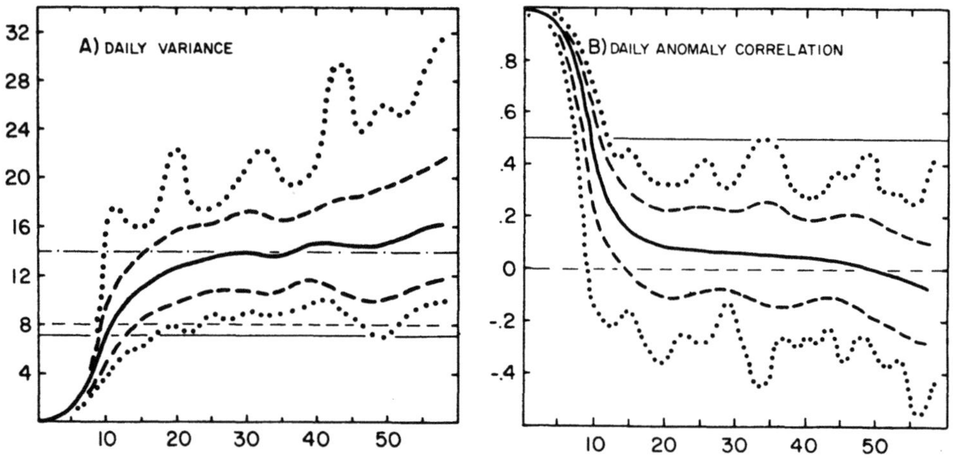

Figure 1. Results of predictability experiments reported in TB for daily 500 mb height fields. Initial perturbation corresponds to a few tenths of a degree in temperature. (a) Error variance in $m^2 \times 10^3$ as a function of forecast time in days for unaveraged forecasts. Heavy solid: ensemble mean, dashed: mean plus or minus standard deviation, dotted: best/worst forecast at time t, light solid: model daily climatological variance, light dashed: observed daily climate variance, light dashed-dot: twice model daily climate variance. (b) Anomaly correlation score as a function of forecast time for unaveraged forecast. Heavy solid: ensemble mean, dashed: mean plus or minus standard deviation, dotted: best/worst forecast at time t, light solid: anomaly correlation of 0.5, light dashed: anomaly correlation of 0.

For forecasts longer than two weeks in advance, detailed weather prediction must be abandoned and the objective of the forecast problem must be redefined. For social and economic reasons the time scale associated with extended-range forecasts of the atmosphere has been one month in advance. This has the advantage of being far enough removed from the medium range so that a clear case can be made for the abandonment of deterministic prediction methods. The traditional objective of such forecasts has been the climate anomaly associated with the mean weather for a particular thirty day period. The meteorological phenomena associated with such anomalies have been variously denoted by the terms regimes, planetary waves and Grosswetterlagen, all referring to the background flow upon which the higher frequency transient eddies are superimposed. Additionally, the one month or extended range prediction must forecast the onset, occurrence and/or decay of long-lived synoptic scale structures (blocks and cut-off cyclones) which can dominate the short term climatology over the time frame of thirty days. From a dynamical systems perspective one must now take a broader, global view of the problem and consider the phase space attractor of the atmosphere (for fixed external forcing). The regimes and long-lived entities described above correspond to inhomogeneities of the density of states on this attractor. Classification of regimes in this manner has been substantiated by Hansen and Sutera (1986) for bimodality of the probability density which is associated with blocking and evidence of multimodality in EOF patterns has been given by Mo and Ghil (1988) and Molteni *et al.* (1990). The latter studies have shown that these regions of (relatively) high probability correspond to the teleconnection patterns delineated by Wallace and Gutzler (1981). Within this context, the extended range prediction problem is that of forecasting in the reduced phase space of regimes and determining the likelihood of maintenance of the current regime and the probability of transition to another regime, *i.e.*, how long the current weather regime will last and what regime will replace the current pattern.

Lastly, the seasonal and interannual prediction problems require that the influence of external boundary conditions be significant in order to maintain a signal in midlatitude weather. The discriminating aspect of atmospheric phenomena with memory over this time scale is that not only is boundary forcing important but the temporal variation of the boundary condition also must be accounted for in order to ensure accurate prediction. The atmospheric phenomena that are predicted in this range are the same regimes (teleconnection patterns) noted above but the distinguishing characteristic is that these patterns are maintained through the action of anomalous boundary forcing and its temporal variation over this time period. Thus the influence of ENSO, midlatitude SST and soil moisture anomalies and sea-ice anomalies and their slow variation in time in altering the lifetimes and amplitudes of the various regimes is the goal of such predictions. From the dynamical systems viewpoint, the prediction problem implies the need to simulate the deformation of the climate attractor due to the anomalous boundary forcing. The prediction is thus the alteration of the probability density of climatic states conditional to the anomalous forcing.

The strength of such changes in probability density are likely to vary significantly depending on latitude—with relatively strong response to local forcing, particularly for tropical SST, and weaker signals relative to noise for remote regions and in general in midlatitudes.

Extended Range Prediction

Because the problem of monthly prediction is the shortest range over which the issues of non-determinism must be addressed, it is natural that the greatest amount of experience has been accumulated for such forecasts. A detailed review of the state of the art of such forecasts is given by Royer (this volume). For the present purposes a few select results will be used to highlight the problems and promises in the extended range.

First, the potential for extended range prediction has been examined by the previously noted studies of Shukla (1981) and Tribbia and Baumhefner (1988). Both these studies showed that at certain times it was possible (in principle) to predict monthly mean anomalies with substantial skill. The study of TB, however, indicated that on average it was not likely that such monthly mean forecasts would be skillful (Figure 2). The expectations of the TB predictability experiments were borne out in the forecast experiments reported on in Baumhefner *et al.* (1989) which demonstrated the lack of consistent forecast skill along with a moderate percentage of skillful forecasts.

In a further analysis of the TB results, Tribbia (1988) noted that much of the skill apparent at the monthly time scale was associated with the leading EOF patterns of monthly mean variability. Together the two most prominent EOF's were responsible for over forty percent of the variance of monthly mean 500 mb geopotential height anomalies and were highly predictable at the thirty day time range. The implications of the model twin experiment results were that the two most prominent regimes of the climate simulation and the transitions between these regimes were predictable events. The verification of this result in forecast experiments has been clouded by the fact that a substantial percentage of errors in the model climate mean projects onto these leading patterns.

However, preliminary indications are that this results carries over from model twin framework to forecast experiments (Branstator, personal communication).

The above theoretical and practical results serve to motivate the necessity of posing the extended range forecast problem in a statistical dynamical framework. Since the forecasts are occasionally accurate and there is apparently a predictable signal embedded in the dynamical noise there is a reasonable chance of useful predictions ensuing from such mixed statistical dynamical prediction methods. The groundwork for the application of such methods has been laid in the works of Epstein (1969) and Leith (1974) and as noted by Royer (loc. cit.), all current efforts at extended range prediction are using a variant of the Monte-Carlo formulation of stochastic-dynamic prediction originally proposed by Leith (1974). The rationale for the use of these methods is to directly capitalize upon the intermittency of skill and the embedded signal aspects of the extended range prediction problem.

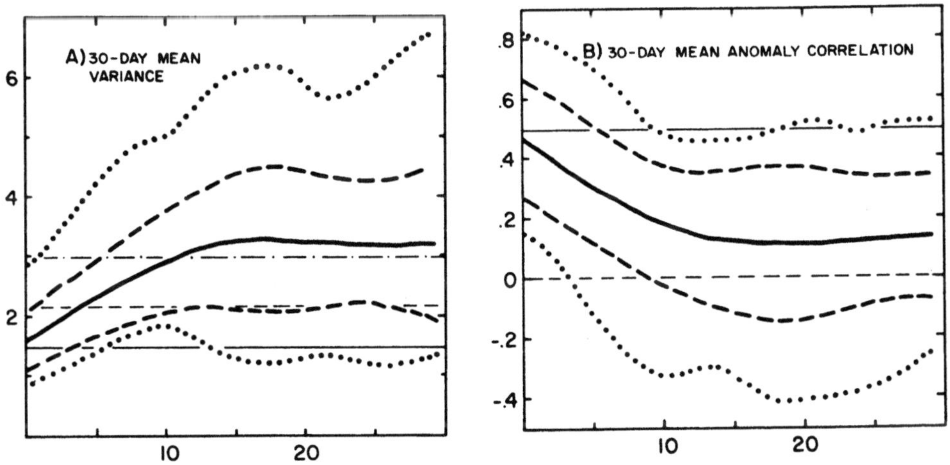

Figure 2a, b. As in Figure 1 for 30-day running means.

In order to capture embedded predictable signals, the ensemble mean of a Monte-Carlo ensemble is used as the optimal (in a Bayesian sense) prediction. This averaging of forecasts can be looked upon as an optimal adaptive smoother which filters the unpredictable noise of the forecasts leaving only the consistent predictable signal. As a filter, it has been demonstrably effective in improving the forecast skill in forecasts with and without external boundary forcing (Baumhefner *et al.* 1989). With regard to the problem of intermittency of skill, the variance structure of the Monte-Carlo ensemble, in principle, should estimate the reliability of the ensemble mean forecast elements through the application of a signal to noise analysis. To date such analyses have been singularly unsuccessful in demonstrating any correlation between predictive skill and forecasted variance (Figure 3). Since the size of typical Monte-Carlo ensembles has been limited to order 10-30, a likely source of the difficulties encountered is undersampling of the underlying (implied) probability distribution. (n.b. Leith estimated that ten ensemble members would be sufficient to realize the dynamical smoothing benefits of stochastic-dynamic prediction. No estimates have been made as to the sampling requisite to describe with accuracy the variance structure of the underlying probability distribution.)

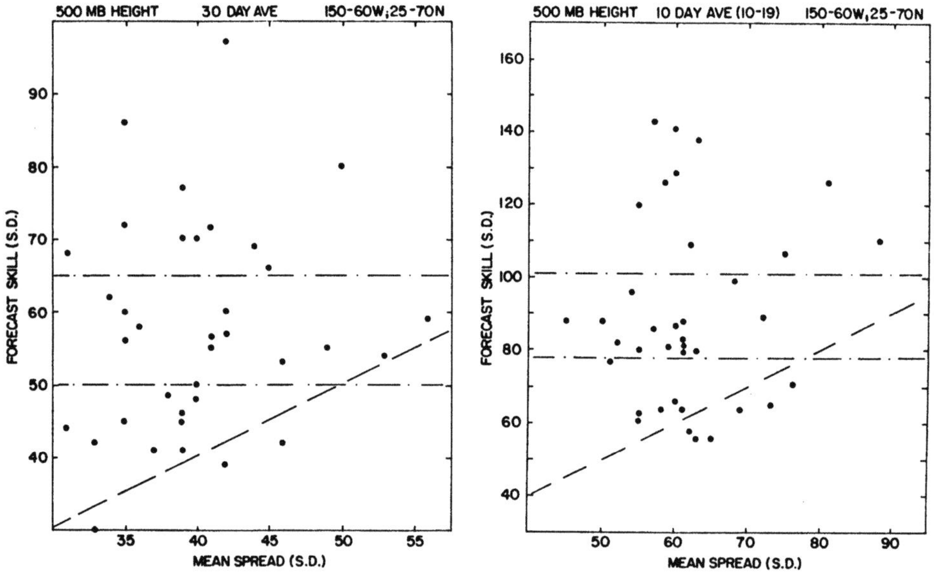

Figure 3. Skill of average forecast from 10-member ensemble in terms of square root of error variance plotted against the standard deviation of the spread among forecasts in the ensemble. 30-day average values on left, 10-day average centered at day 14.5 on right (from Baumhefner, 1989).

Prediction of Reliability

Because the difficulties in predicting reliability become evident even in the short and medium range, it is necessary to consider the problem in these early stages. This is due to the fact that the loss of skill in a probabilistic forecast of reliability in the near term will likely preclude skill in the extended range. As will be seen below, this simplifies the problem to some degree and will allow a quasi-analytic treatment of the the issues surrounding the prediction of reliabilty.

It is particularly useful to consider the ideal prediction problem of the stochastic-dynamic techniques at this juncture since the simplifications and restrictions necessary for practical implementation are potentially a source of the difficulties heretofore encountered. As intimated in the previous section, the proper prediction problem for statistical-dynamic problems is the evolution in time of the underlying probability density. Such predictions have been considered in the context of classical statistical mechanics where the theory has been developed by Gibbs and Liouville. The prognostic equation for the probablity density of a dynamical system governed by

$$\dot{\mathbf{x}} = \mathbf{f}(\mathbf{x}) \tag{4.1}$$

is simply the continuity equation for the density, p, with 'fluid velocity' given by equation (4.1) above; *i.e.*,

$$p_t + \frac{\partial}{\partial \mathbf{x}} (\mathbf{f} p) = 0 \tag{4.2}$$

This is a simple reflection of the fact that probability can neither be created nor destroyed identical to mass in a mechanical context.

For the problem of forecast reliability one is most interested in the probability centered about a particular forecast, thus it is useful to transform the prognostic system for individual forecasts to a prognostic system for deviations about a control trajectory $\mathbf{x}_c(t)$. Defining $\mathbf{y} = \mathbf{x} - \mathbf{x}_c$, the dynamical system in terms of these deviatoric coordinates is

$$\dot{\mathbf{y}} = \mathbf{f}(\mathbf{x}_c + \mathbf{y}) - \mathbf{f}(\mathbf{x}_c) \tag{4.3}$$

Since the short time behavior of the above system is of most immediate interest, linearization of (4.3) is justifiable and (4.3) can be simplified to

$$\dot{\mathbf{y}} = \mathbf{A}(\mathbf{x}_c(t))\, \mathbf{y} \tag{4.4}$$

where

$$\mathbf{A} = \left.\frac{\partial \mathbf{f}}{\partial \mathbf{x}}\right)_{\mathbf{x}=\mathbf{x}_c}$$

The Liouville equation for the dynamical system in terms of the variable $\mathbf{y}$ is then

$$p_t + \frac{\partial}{\partial \mathbf{y}} (\mathbf{A}\,\mathbf{y} p) = 0$$

or

$$p_t + \sigma(t)p + \mathbf{A}\mathbf{y}\frac{\partial p}{\partial \mathbf{y}} = 0 \tag{4.5}$$

where $\sigma(t)$ = Trace **A**. With $\phi = e^{\int_0^t \sigma(s)ds} p$, (4.5) can be rewritten in the following 'advective' form:

$$\phi_t + \mathbf{V} \cdot \nabla_y \phi = 0 \tag{4.6}$$

where $\mathbf{V} = \mathbf{A}\ \mathbf{y}$.

In the solution of (4.6) the natural technique is the method of characteristics. This method utilizes the constancy of ϕ along the directions in spacetime traced by the integral curves of (4.4). As demonstrated in the appendix, if at $t = 0$ the probability density is a homogeneous function of $\mathbf{y}$; *i.e.*, $p(\mathbf{y}, 0) = \mathbf{F}(\mathbf{y}^+\mathbf{S}^+\mathbf{S}\,\mathbf{y})$, with **S** a symmetric positive matrix and $\mathbf{S}^+$ its adjoint, then it is a straightforward calculation to show that $\phi(\mathbf{y}, t)$ is given by

$$\phi(\mathbf{y}, t) = \mathbf{F}\left(\mathbf{y}^+\mathbf{R}_I^+\mathbf{S}^+\mathbf{S}\ \mathbf{R}_I\mathbf{y}\right) , \tag{4.7}$$

where $\mathbf{R}_I$ is the inverse of the resolvent **R** defined in the appendix.

The above analytical solution can elucidate the nature of the current problems in attempting to predict forecast reliability. First, because the method of characteristics is formally identical to the procedure of Monte-Carlo prediction, the resolution of our current difficulties can be gained by requiring that our Monte-Carlo method mimic important aspects of the method of characteristics. Next, our understanding of the issues can be enhanced through the exploitation of the fluid dynamic analogy inherent in equation (4.6). In the estimation of forecast reliability it is likely that the extremes of the probability density distribution in phase space are requisite; *i.e.*, the domains in phase space which have large relative probability and are distant from the centroid of the distribution. The solution (4.7) determines the radial directions from the center with maximum probability as the direction parallel to the eigenvector of

$$\mathbf{M} = \mathbf{Q}^+\mathbf{R}^+\mathbf{R}\mathbf{Q} \tag{4.8}$$

with largest eigenvalue (cf. appendix).

The directions associated with the largest eigenvectors thus correspond to the most dangerous or most sensitive directions with respect to the prediction of the probability density and clearly the reliability prediction problem is intertwined with the dynamics of these dynamical degrees of freedom. Because of the fluid dynamic analogy, a kinematic interpretation can be arrived at for these directions (under the special circumstances of uniform, diagonal **S** and temporally invariant **A**). With these restrictions, the eigenvectors corrrespond to the principal axes of strain and the eigenvalues the principal rates of strain for the linear velocity field $\mathbf{V} = \mathbf{Ay}$. And, just as fluid line elements are stretched most rapidly if they are aligned to a direction associated with a large principal rate of strain, the probability density ϕ is also stretched and strained most efficiently in these particular directions. Thus, these are the directions along which a forecast potentially suffers the greatest loss of skill.

With these preferred directions in mind, the adequacy of the random sampling techniques used currently (Royer op. cit) can be quantitatively evaluated. The ability of any Monte-Carlo forecasting system to produce accurate reliability predictions in the short term can be reduced to the ability of the initial sampling methodology to span the most sensitive directions in phase space as defined by the eigenmodes of (4.8) in a representative manner.

It must be noted at this juncture that because the dimensionality of the eigenproblem is identical to that of the dynamical system which corresponds to an order of magnitude of 10^6 for a realistic forecast model of the atmosphere, the solution of the eigenvalue problem (4.8) is impractical at the current time. However, one can get a feel for the spectrum and characteristic spatial structures in such a problem by using a simplified model of the atmosphere. For the purpose of investigating the sampling issues a two-level quasi-geostrophic model (T31 spectral horizontal resolution) will be used. To further simplify matters the interval of time over which **R** is computed will be a single time step. Thus, the resulting eigenproblem concerns finding the principal axes and rates of strain for $\mathbf{A}(0)$ or the eigenvectors and eigenvalues of

$$\mathbf{A} + \mathbf{A}^{+}$$

with a neglect of terms of order (Δt). Such simplifications are not germane to the problem of sampling and *a posteriori* can be shown to have negligible effect on the results. The eigenvalue spectrum obtained from this simplified problem is shown in Figure 4.

In order to examine the efficacy of random sampling a computationally constructed distribution consisting of 2000 realizations of random vorticity deviation fields was generated with a realistic spectral distribution of random error potential vorticity variance. Specifically, a white noise quasi-geostrophic potential enstrophy spectrum was used with amplitude such that two-dimensional wavenumber 31 was typical of the observed potential enstrophy in this scale. This grand ensemble was broken into sub-ensembles of varying size (1-30 elements) typical of current 'operational' Monte-Carlo forecasting practice.

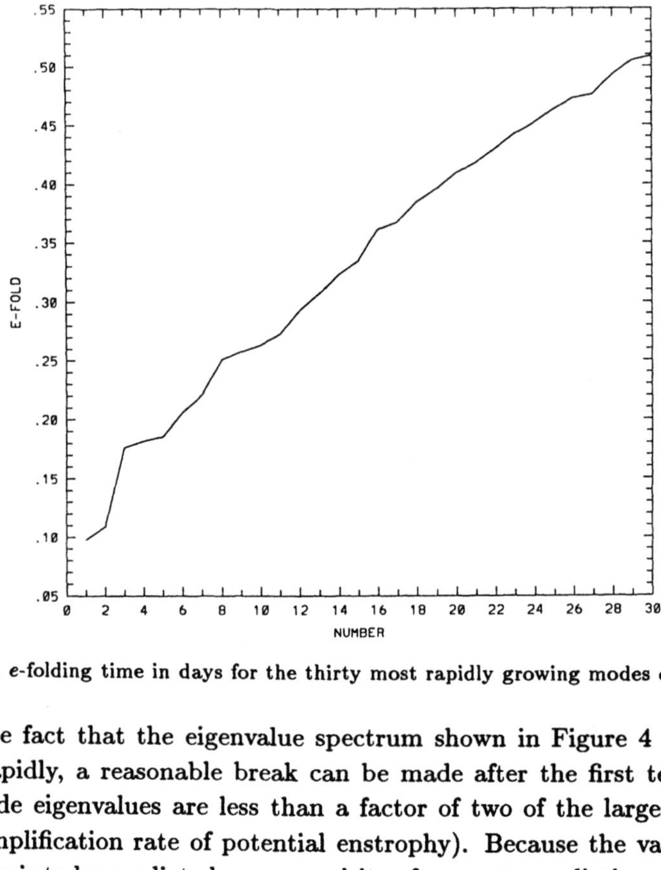

Figure 4. *e*-folding time in days for the thirty most rapidly growing modes of $\mathbf{A} + \mathbf{A}^+$.

Despite the fact that the eigenvalue spectrum shown in Figure 4 does not fall off very rapidly, a reasonable break can be made after the first ten where the higher mode eigenvalues are less than a factor of two of the largest eigenvalue (largest amplification rate of potential enstrophy). Because the variance of the distribution is to be predicted, a prerequisite of accurate prediction of this second moment quantity is that the initial sampling strategy gives a representative initial amplitude to the first ten (1-10) eigenmodes corresponding to the ten most rapidly amplifying directions in phase space. A representative amplitude for the present purposes will be defined as being within fifty percent of the root mean square amplitude computed from the underlying probability distribution of two thousand realizations. With such a criterion for the accuracy of the initial second moments of a Monte-Carlo ensemble, the effect of finite sampling at various sample sizes can be quantitatively evaluated. Figure 5a shows the estimated probability of any one of the most amplifying modes receiving a representative initial amplitude for an ensemble size of ten. As can be seen, ensembles of ten or more elements have a high probability of one of the most dangerous modes receiving an accurate initial amplitude. However, a more stringent and important criterion is that all ten of the fastest amplifying modes have representative initial values. As can be seen in Figure 5b only one out five such ensembles with ten elements satisfy this stricter criterion. With the relatively small probabilities associated with even the rather limited criteria of ten eigenmodes having less than 50% error in their amplitude, it is clear that sampling inadequacies of the random sampling technique can easily be at the root of the reliability prediction problems encountered to date.

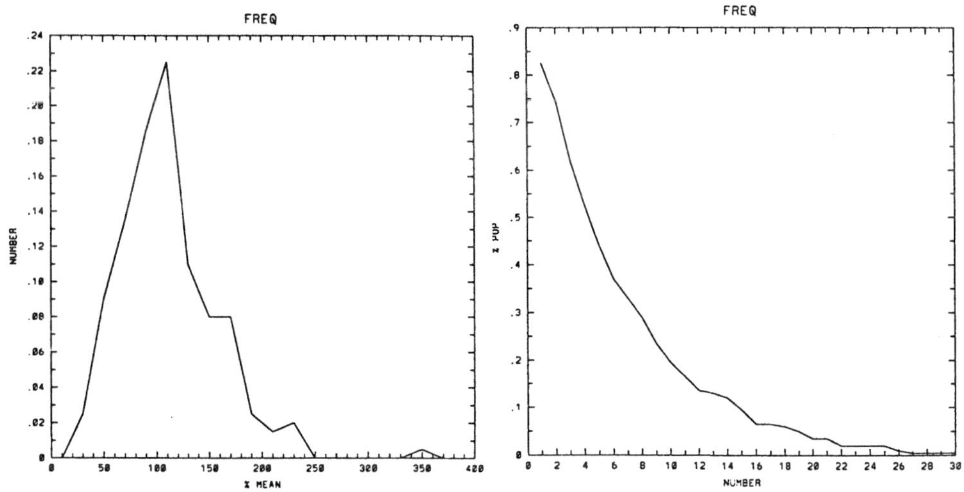

Figure 5. a) Probability density of the most unstable mode having an ensemble mean variance relative to the true variance for an ensemble of size ten. b) Probability of having representative variance values for modes up to and including mode M for a ten member ensemble.

It must also be noted here that another potential cause of the poor result shown in Figure 3 is that random modeling errors (Tribbia and Baumhefner (II), 1988) are the dominant cause of forecast errors. However, the fact that the inaccuracies in forecasting reliability arise very early in the forecast period (within the first three days) argues against modeling error dominance (TB II). Furthermore, the results shown above demonstrate the inadequacies of small random samples for skill prediction even without the influence of model errors. In addition to the random sampling strategy, a variant which has been used in the context of operational prediction centers where forecasts initiating at different times have been used as an ensemble for a common prediction time is known as lagged averaging (Hoffman and Kalnay, 1983). This technique has advantages and disadvantages relative to random sampling. Although it is limited to small sample sizes and thus sampling deficiencies, it has the advantage that in data-rich areas it contains an estimate of the total error in the forecast system at the time of the initiation of the most recent forecast in the ensemble. On the negative side for this technique is the difficulty that in regions of little information input into the initial analysis members of the ensemble will tend to differ little because of the use of the previous forecast in the data-assimilation. Thus, there will be an underestimate of the uncertainty in these regions of little data leading to an optimistic prediction of forecast reliability. It is likely that for the small sample sizes necessitated by the need for contemporaneous forecast date for the entire ensemble the benefits and disadvantages of this method will act to give skill in reliability predictions similar to those obtained via random sampling for the same sample size.

A Rational Monte-Carlo Method

From the above analysis, a practical and rational scheme for Monte-Carlo sampling can be developed. The methodology that is implied by the above is to construct the Monte-Carlo ensemble by projecting an estimate of analysis error variance onto the N-most unstable directions in phase space, where N is a number of order 20-50. Such a scheme seems assured of success in the short range, as experimentation at the European Centre, KNMI and NCAR is currently tending to demonstrate. The major question that arises is the adequacy of such sampling methods in the extended range where multimodality of the probability density is of prime importance. The problems associated with the prediction of such bifurcations of the distribution will be the next test for this technique and will be the critical test for the utility of small sample size ensemble integrations.

Acknowledgements

We would like to thank Ronna Bailey for her efforts in the production of this manuscript and note the fruitful collaboration with our colleagues at NCAR, in particular Grant Branstator, and at the European Centre for Medium-Range Weather Forecasts. Lastly, JJT wishes to dedicate this work to the memory of his father, Luke Tribbia, who passed away during the initial stages of this study. He has missed his guidance and love in the intervening time.

References

Baumhefner DP (1989) Prediction of forecast skill using Monte-Carlo ensembles of 30-day forecasts. In: Proceedings of the Fourteenth Annual Climate Diagnostics Workshop, La Jolla, CA, 16-20 October 1989, Climate Analysis Center/National Meteorological Center, Washington, D.C., pp. 339–343.

Baumhefner DP, Tribbia JJ, Blackmon ML (1989) The influence of specified SST and initial condition uncertainty on Monte-Carlo extended range ensembles. In: Proceedings of the Workshop on Intercomparison of Results from Atmosphere and Ocean General Circulation Models, Technical Document No. 254. World Meteorological Organization, Geneva.

Epstein ES (1969) Stochastic dynamic prediction. Tellus 11:739-759

Hansen AR, Sutera A (1986) On the probability density distributions of planetary-scale wave amplitudes. J Atmos Sci 43:3250-3265

Hoffman RN, Kalnay E (1983) Lagged average forecasting, an alternative to Monte Carlo forecasting. Tellus 35A:100-118

Leith CE (1971) Atmospheric predictability and two dimensional turbulence. J Atmos Sci 28:145-161

Leith CE (1974) Theoretical skill of Monte-Carlo forecasts. Mon Wea Rev 102:409-418

Lorenz EN (1969) The predictability of a flow which possesses many scales of motion. Tellus 21:289-307

Mo K, Ghil M (1988) Cluster analyses of multiple planetary flow regimes. J Geophy Res 93(D): 10,927-10,952

Molteni F, Tibaldi S, Palmer TN (1990) Regimes in the wintertime circulation over northern extratropics. I: Observational evidence. Quart J Roy Met Soc 90A:31-68

Shukla J (1981) Dynamic predictability of monthly means. J Atmos Sci 38: 2547-2572

Smagorinsky J (1969) Problems and promises of deterministic extended range forecasts. Bull Amer Met Soc 50:286-311

Thompson PD (1957) Uncertainty of initial state as a factor in the predictability of large scale atmospheric flow pattern. Tellus 9:275-295

Tribbia JJ (1988) Predictability of monthly mean teleconnection patterns. Pont. Acad. Scien. Seripta Varia 69:567-592

Tribbia JJ, Baumhefner DP (1988) Estimates of predictability of low-frequency variability with a spectral general circulation mode. J Atmos Sci 45:2306-2317

Tribbia JJ, Baumhefner DP (1988) The reliability of improvements in deterministic numerical forecasts. Mon Wea Rev 116:2276-2288
Wallace JM, Gutzler DS (1981) Teleconnections in the geopotential height field during the Northern Hemisphere winter. Mon Wea Rev 109:784-812

APPENDIX

Equation (4.6) can be solved using the method of characteristics in the following manner. With

$$\phi_t + \mathbf{A}\,\mathbf{y}\,\frac{\partial \phi}{\partial \mathbf{y}} = 0$$

and

$$\phi\,(\mathbf{y}, 0) = F\left(\mathbf{y}^{+}\mathbf{S}^{+}\mathbf{S}\,\mathbf{y}\right)\ ,$$

the method of characteristics states that $\dfrac{d\phi}{ds} = 0$, where s is defined by the space-time curves which solve

$$\frac{d\mathbf{y}}{ds} = \mathbf{A}\,(t)\,\mathbf{y} \qquad \text{and} \qquad \frac{dt}{ds} = 0\ .$$

The definition of the resolvent, $\mathbf{R}$, for the linear system

$$\frac{d\mathbf{y}}{ds} = \mathbf{A}\,(s)\,\mathbf{y}$$

is that linear transformation which, for any initial vector $\mathbf{y}_0$, gives upon multiplication by $\mathbf{R}$ the solution to the linear system at time t, *i.e.*,

$$\mathbf{y}\,(t) = \mathbf{R}\,(t, 0)\,\mathbf{y}_0\ .$$

It is straightforward to show then that

$$\phi\,(\mathbf{y}, t) = F\left(\mathbf{y}^{+}\,\mathbf{R}_I^{+}\,\mathbf{S}^{+}\,\mathbf{S}\,\mathbf{R}_I\,\mathbf{y}\right)$$

where $\mathbf{R}_I = \mathbf{R}^{-1}\,(t, 0)$. Now the maximization of $\phi\,(y, t)$ along a radial arm corresponds to the minimization of

$$\mathbf{y}^{+}\,\mathbf{R}_I^{+}\,\mathbf{S}^{+}\,\mathbf{S}\,\mathbf{R}_I\,\mathbf{y}$$

with the constraint $\mathbf{y}^{+}\,\mathbf{y} = 1$.

This variational problem can be re-cast using the resolvent, $\mathbf{R}$, rather than its inverse, giving a maximization of

$$\mathbf{y}^{+}\,\mathbf{Q}^{+}\,\mathbf{R}^{+}\,\mathbf{R}\,\mathbf{Q}\,\mathbf{y}$$

with $\mathbf{y}^{+}\,\mathbf{y} = 1$, where $\mathbf{Q} = \mathbf{S}^{-1}$. Defining the symmetric matrix $\mathbf{M}$ as $\mathbf{Q}^{+}\,\mathbf{R}^{+}\,\mathbf{R}\,\mathbf{Q}$, this maximization corresponds through the Rayleigh-Ritz property of the symmetric eigenvalue problem to finding the largest eigenvalue and corresponding eigenvector of $\mathbf{M}$.

GPSR Compliance

The European Union's (EU) General Product Safety Regulation (GPSR) is a set of rules that requires consumer products to be safe and our obligations to ensure this.

If you have any concerns about our products, you can contact us on ProductSafety@springernature.com

In case Publisher is established outside the EU, the EU authorized representative is:

Springer Nature Customer Service Center GmbH
Europaplatz 3
69115 Heidelberg, Germany

Batch number: 10620878

Printed by Printforce, the Netherlands